RENAISSANCE COMBAT

RENAISSANCE COMBAT

Jörg Wilhalm's Fight Book, 1522–1523

DIERK HAGEDORN

with HELEN & HENRI HAGEDORN

Foreword by TOBIAS CAPWELL

†

GREENHILL BOOKS

Published in 2021 by Greenhill Books,
c/o Pen & Sword Books Ltd,
47 Church Street, Barnsley, S. Yorkshire, S70 2AS
www.greenhillbooks.com

ISBN 978-1-78438-656-6

Library of Congress Cataloging-in-Publication data available

Typeset in Charpentier Renaissance by Dierk Hagedorn
Images used by kind permission of
Bayerische Staatsbibliothek München, Cgm 3711

Printed and bound in India by Replika Press Pvt. Ltd.

CONTENTS

Tobias Capwell

FOREWORD

For hundreds of years, an entire genre of medieval and Renaissance European literature and visual art has remained largely unnoticed. It is a fabulously rich and immediately revealing body of work tracing its origins back to the late thirteenth century (at least) and extending through the late medieval period, through the Renaissance, and well into the modern era. It is comprised of thousands of unique manuscripts and early printed books. While these enigmatic volumes were never completely unknown, their early cataloguers in libraries, archives and museums tended not to know what to make of them.

The 'fight book' is unlike most other books. It cannot simply be read and understood. Its use requires not only exceptional scholarly talents – fluency in old languages and mastery of the art of palaeography – but also considerable experience with traditional hand weapons and their use. Not only does the reader need to know the esoteric terminology of fighting, he/she has to have a personal, practical understanding of what being in a fight feels like – how the speed of the action can suddenly change, how the opponent will try to control your state of mind, how the precise length and direction of a step or the angle of an attack can mean victory or defeat. The fight book tries to convey practical ideas to literate martial artists, thoughts which are intended to soak into the mind of the reader and thus assist his/her physical practice. Without that practice however, the fight book is largely useless. No one ever became a great sword fighter by reading books.

The fight book is not meant to exist or be used in isolation; it is intended to work as part of a very specific environment, a place where the ability to fight well with the sword, dagger, axe and spear was a vital life skill, where being trained and ready to confront one's enemies in full plate armour could earn one money, fame and prestige. Again, these skills could not be found in the pages of any book, but only through rigorous, dedicated, single-minded practice. Maintaining the necessary mental focus is as difficult, if not more difficult, than enduring the required physical exertion and punishment. This may be where the fight book finds its place in the world of combat.

When I look at a fight book, especially one – such as that of Jörg Wilhalm – produced during the German Renaissance, I feel many things. As an art historian I am immediately drawn in by the book's value as a work of art, a visual document of aspects of the material culture of its time. As an arms and armour specialist, I am pulled deeper by the way it hints at the original, living context of objects which I normally study in the disassociated setting of the library, museum and laboratory. Finally, and perhaps most significantly, as someone who has practised martial arts since childhood, the fight book makes me want to be a better fighter. It inspires me with a renewed sense that however much I have learnt, however well I may be able to defend myself, there is always more – more technique, more variation, more subtlety. The fight book is a spur to physical practice.

For these reasons, it should come as no surprise that historically, most curators, scholars and archivists have had no idea what to do with this material. We should not blame them. Working with this material is an enormous challenge, requiring a combination of key skills and knowledge that have not been employed together for hundreds of years. The person who wrote the fight book, and the original intended readership, regarded letters and combat to be equal parts in a respectable education. Ultimately, this remains the only way to truly understand the fight book.

Over the last few decades, I have watched with enduring fascination as the modern historical European martial arts community has grown, proliferated and thrived. When I looked upon an original fight book for the first time (the now world-famous I.33, the earliest known fencing text), sitting in the library of the Royal Armouries in the Tower of London almost thirty years ago, I could never have imagined that these books would soon inspire thousands of people across the world to dedicate significant portions of their lives to the study and practice of these ancient fighting systems. It is a remarkable culture, one that has expanded immeasurably our understanding of the real human experience of life in the past.

As already noted, the interpretation of these illustrated texts requires special academic and practical gifts. When I first met the author of the present volume, it was not in an archive. I was standing outside the walls of a fortified manor house in central Germany, clad in full plate armour and armed with a pollaxe. I faced another similarly equipped person, whom I had never met. Indeed, the only hints of his identity were the round-rimmed spectacles peeking out of the gap between sallet and bevor. Dierk Hagedorn is the very embodiment of the scholar-swordsman. After having accepted the honour of being hit by him, I was pleased to attend one of his practical clinics, where he discussed the contents of a German longsword treatise with the ease of true authority, and with, literally, a sword in one hand and a book in the other.

Jörg Wilhalm and the reader are in good hands.

Tobias Capwell is Curator of Arms and Armour at the Wallace Collection in London.

INTRODUCTION

'Still sharp.' BOROMIR, *The Lord of the Rings: The Fellowship of the Ring*, 2001

I came to realise that Jörg Wilhalm can still be dangerous today, although he's been dead for almost five centuries: when I pulled some printouts used in the preparation of this edition out of the printer, I cut myself on a sharp edge of the paper. I bled only a little. The combatants shown in the marvellous illustrations of Jörg Wilhalm's codex may not have come out of their encounters so easily.

Books that feature single combat techniques have a long tradition. *Fechtbücher*, as they are called (in English: fight books), were produced beginning in the 14th century in the form of manuscripts, the majority of them in Germany. Later ones were printed, but even after the advent of print in the year of 1455, many fight books were entirely hand-crafted. This is the case for the five volumes attributed to Jörg Wilhalm, which stem from the 1520s. At that time, the kind of serious combat they present, the duel in and out of armour, with the long sword and other medieval weapons, was already falling out of fashion as more modern weapons arose, particularly the rapier but also firearms. In this regard, Jörg Wilhalm was one of the final witnesses of an ancient art that would soon be looked at as only a curiosity, as a distant relic from the past. Wilhalm was not only a practitioner but also a preserver of an art that was slowly falling into oblivion.

The last ten or twenty years, however, have seen a rediscovery, a resurrection of this lost art, resulting in a number of publications about centuries-old fight books, bringing them back from the shadowy vaults of libraries and museums where they had been silently resting and presenting them to an eager, modern audience. Some of these editions offer a general overview of the matter, others examine only one single source – as this one does.

Jörg Wilhalm has never received much scholarly attention despite the number of manuscripts that are connected to his name. A manuscript from Munich, Codex germanicus monacensis 3711, is an extensive collection of old fighting techniques and contains material possibly from Wilhalm's own hand, or – more likely – material commissioned by him. It consists of illustrated techniques on sword fighting, two versions of fighting in armour, combat on horseback, and a short section of miscellanea. In addition, it features an extensive text – not illustrated – about the long sword in the tradition of the famous grandmaster Johannes Liechtenauer.

I am eternally grateful for the invaluable assistance of my two children Helen and Henri, who were 14 and 11 years old when the project commenced. Their help in this endeavour was considerable, a true asset I would not have thought to be possible.

The whole idea started as a joke. After my previous book had gone to the printers, I told my children that they possibly should make a little effort too; after all, they had not yet published anything. How would they like to return to school from their next holidays and, upon being asked by their teachers what they had done in their spare time, they could answer: 'We've made a book.' 'Ha-ha,' the teachers would reply; and 'Ta-daa,' my children would answer in return, pulling printed books out of their schoolbags with their names on them – on the books, not the schoolbags.

Of course, making a book takes considerably longer than just a few weeks of holiday, but nevertheless, the idea stuck and they insisted on actually helping.

And so it began.

conquer Constantinople (modern-day İstanbul), thereby ending the Byzantine (or Eastern Roman) Empi

My daughter Helen sat next to me time and again in order to tackle the intricacies of Early New High German handwriting. I owe her the transcriptions of the illustrated parts of the manuscript on sword fighting, harness fencing and mounted combat. I merely contributed the text-only passages about fighting with the long sword.

Meanwhile, my son Henri almost single-handedly compared the enormous number of images from 15 manuscripts that were used for the concordance; see page 394.

In a family effort, we were able to bring one of the most important manuscripts about fighting techniques in the early Renaissance in Germany to life. For the very first time, Wilhalm's work appears in print in all its colourful splendour. Furthermore, the Codex germanicus monacensis 3711 must be considered the most compact and comprehensive one of the five manuscripts that are attributed to Jörg Wilhalm, hatter in Augsburg.

DIERK HAGEDORN

ter more than a thousand years. 1455: Johannes Gutenberg completes his edition of the Bible, printed with

Dierk Hagedorn

Jörg Wilhalm – Fight like a Hatter

Editorial History

This volume is the first edition about the manuscript Cgm 3711 – and about Jörg Wilhalm in general. Previously, it has only been mentioned ever so briefly in the specific literature.[1] In the 19th century, the German philologist J. A. Schmeller[2] listed the manuscripts of Munich's principal library, at the time called Königliche Hof- und Staatsbibliothek, without any further information than the shelf mark, year of origin, size, author and title – *Fechtkunst* ('Art of Fighting') in this instance. Roughly 100 years later, Martin Wierschin[3] gave a brief outline of the contents and general character of the manuscript with four black and white reproductions in an appendix. Another 20 years later, Hans-Peter Hils[4] repeated Wierschin's account more or less unaltered but also provided us with additional information. He claimed that one part of our manuscript was a copy of two manuscripts from Augsburg (JWA2 and JWA3 – for scribal abbrevia-

1 In 2017, however, the reprint publishing house Fines Mundi issued a heavily colour-corrected reproduction as *Die Fechtkunst von Jörg Wilhalm Huter*. The volume offers no further information apart from the images. Curiously, the edition notes the original as being published in 1523 – notwithstanding the circumstance that manuscripts, objects that *per definitionem* are unique items, are not exactly 'published'.

2 Schmeller 1866, p. 386.

3 Wierschin 1965, p. 29, images 35–38.

4 Hils 1985, pp. 98–99; pp. 192–195.

tions, see below, 'Sources used') and is in return the source for yet another manuscript from Augsburg, JWA1.[5] These assertions, however, could not be confirmed. The catalogue of German illustrated medieval manuscripts, published in 2008, offers far more insight.[6] The following codicological information owes a small part to this catalogue, but since its description is flawed in many respects, I present a corrected and heavily augmented version, incorporating information provided by the Bayerische Staatsbibliothek, the institution in possession of this valuable manuscript.

THE CODEX

The Codex germanicus monacensis 3711 was produced in Augsburg and is dated internally in four places: 1522 (fol. 59r) and 1523 (fol. 42r, 45r and 51r). The language is Swabian.

Provenance

The manuscript was either written or commissioned by Jörg Wilhalm, who worked as a hatter in Augsburg, in 1522 or 1523 as these dates appear on the pages. Later, the manuscript came into the possession of the Jesuit college in Augsburg, founded in 1582. There is a note to this effect on the first page of the manuscript: Nr° 230 Bibl. Jesuit. A. Vind. 53/4. It was transferred to the Königliche Hof- und Staatsbibliothek zu München (Royal Court and State Library in Munich) in 1807 by Bernhard Joseph Docen. The library would later become the Bayerische Staatsbibliothek, where the manuscript is preserved to this day.

5 Hils 1985, p. 40.
6 Leng 2008, p. 85–87.

Contents

Ir–XIXv	38 blank pages.
1r	An abridged and modified version of Johannes Liechtenauer's introduction to fencing with the long sword. 1v is blank.
2r–42r	Images and captions of unarmoured sword fencers. The captions are partially based on Johannes Liechtenauer's verses and are in other parts (possibly) an invention of Jörg Wilhalm. 42v is blank.
43r–45r	Johannes Liechtenauer: The art of the long sword, verses.
45v–51r	Johannes Liechtenauer: The art of the long sword, verses and glosses. 51v is blank.
51ar–51kv	20 blank pages.
52r–58v	Combat in armour, without text.
58ar/v	Miscellanea: Combat between bearded men with shields and clubs; man against woman; fight with curved knives.
58br–58ccv	54 blank pages.
59r–84r	Combat in armour with captions. 84v is blank.
84ar/v	2 blank pages.
85r–102r	Mounted combat with captions. 102v is blank.
103r–106v	8 blank pages.

Codicological description

The material is paper. 160 sheets.[7] Only the 102 sheets with text or images were foliated in ink, possibly by the writer's hand – 24 was initially erroneously foliated as 23, 26 as 25 and 51 as 49, all of them struck out and corrected; empty pages were not taken into account for the old numbering but have later been foliated in pencil, after the digitisation: 19 sheets before 1, 10 sheets after 51, 27 sheets after 58a, one sheet after 84, four sheets after 102. The top right corner of fol. 2

7 A close inspection verifies this number, in disagreement with Leng, who counts 163 sheets.

is torn, but the foliation is visible and was obviously written after the damage occurred. Size: 310 × 220 mm. The text is written in a Cursive of the 1st quarter of the 16th century with strong reminiscences of older Bastarda, by one hand; writer corresponds to Augsburg, Cod. I.6.2°2 (JWA1) 2r–49r and Cod. I.6.2°3 (JWA2) 1r–42v. One single column of text on 1r with 24 lines, text blocks with up to 35 lines on 43r–51r, otherwise captions of 3–12 lines. No Lombardic capitals, only two initials with simple wickerwork on 26v and 42r. Not rubricated. The book cover consists merely of a sheet of parchment wrapped around and sewn to the body of the book, secured to it by three leather straps on the spine that overlap to the front and back covers where they are inserted in three slits. On the spine, there is a light yellow label with the shelf mark. Another similar label can be found on the top left corner of the inside of the front cover.

Watermarks

There are several old tracings in the watermark information system (Wasserzeichen-Informationssystem – WZIS)[8] which, however, cannot be assigned to individual sheets.

 = ox head
wasserzeichen-online.de/?ref=DE5580-Cgm3711_XXXXXXXX
wasserzeichen-online.de/?ref=DE5580-Cgm3711_XXXXXXX
wasserzeichen-online.de/?ref=DE5580-Cgm3711_XXXXXX
wasserzeichen-online.de/?ref=DE5580-Cgm3711_XXXXX
wasserzeichen-online.de/?ref=DE5580-Cgm3711_XXXX
wasserzeichen-online.de/?ref=DE5580-Cgm3711_X

 = letter P (mostly covered by text or image in the manuscript)
wasserzeichen-online.de/?ref=DE5580-Cgm3711_XX

8 wasserzeichen-online.de.

= pine cone (from the Augsburg city coat of arms)
wasserzeichen-online.de/?ref=DE5580-Cgm3711_XXX

Information provided by the university library in Augsburg reveals that the two sister manuscripts JWA1 and JWA2 share a watermark, an ox head identical to the one in our *Leithandschrift* – the primary source – JWM1. Nevertheless, there is a difference since the ox head in our manuscript appears mostly on the blank sheets. Quire 2, which corresponds to JWA1, primarily contains the P watermark, while quire 5, equivalent to JWA2, bears the pine cone watermark.

Quire analysis

Formula: $\mathrm{VIII}^{\mathrm{XVI}} + (\mathrm{XXVII} + \mathrm{I})^{51b} + \mathrm{XII}^{58i} + \mathrm{IX}^{58bb} + \mathrm{XXV}^{106}$

In the following diagram, blank sheets have a beige background.

Quire 1: VIII

Quire 2: XXVII + I

Upper line: first half of the quire from left to right; lower line: second half of the quire from right to left; i. e. the sheets in the upper and lower line form one double sheet in each case.

Threads are visible after fol. 4, 11 (here, a double sheet was sewn in), 18, 26 (centre of the quire). Foll. 11/12: double sheet inserted into the quire.

Quire 3: XII

Quire 4: IX

Quire 5: XXV

Graphical representation analogous to quire 2.

1493: Maximilian I succeeds his father, Frederick III, as ruler of the Holy Roman Empire. 1498: Vasco d

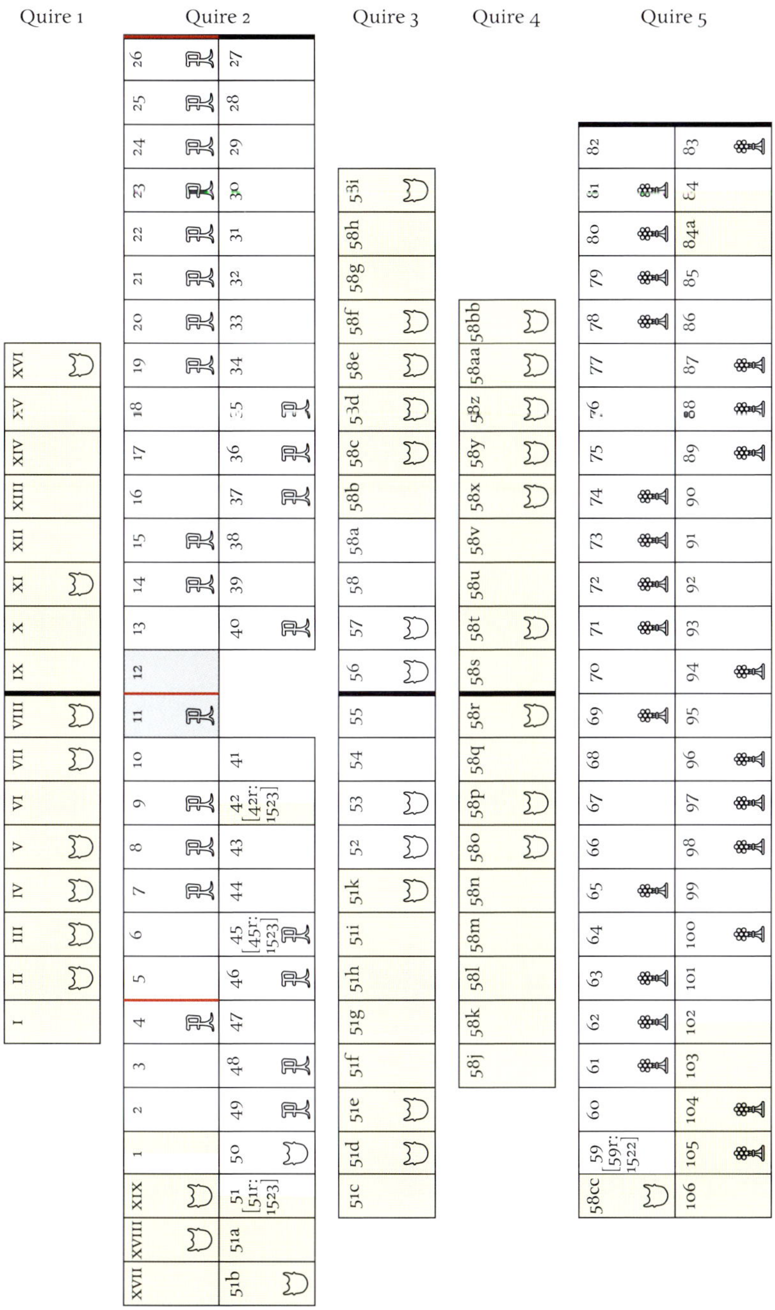

ama discovers the maritime route to India. – The preacher and ruler of Florence, Girolamo Savonarola, is

Format and page layout

Fol. 2r–42r: 150 mm high illustrations on the top of the page, divided by a line from the lower half with explanations. Fol. 43r–51r: text-only pages. Fol. 52r–58av: two 150 mm high half-page illustrations per page (except for fol. 58av, which has only one image), without any accompanying text. Fol. 59r–60r: 230–250 mm high images on the upper half of the page, divided by a line with short text below or within the illustration in a scroll (fol. 59v). Fol. 60v–101v: 160–195 mm high illustrations in the upper half of the page; explanatory texts are situated under a dividing line. Fol. 102r: two 120 mm high images, divided by a line, with short descriptions.

The process by which the manuscript was created is not entirely clear. In all probability, the line drawings of the illustrations were created first, followed by either colouring or writing the text. In some instances, single letters from the lower part of the page reach into the upper part with the illustration (2r, 4v, 61v). But due to the very subtle shade of the washing, it cannot be determined whether the drawings were filled with colour first or whether the text was written beforehand. In any case, the text followed the drawings: the dividing line between image and text avoids the crossguard of a fallen sword on fol. 69v.

The illustrations were in all probability prepared faintly in pencil; numerous very imprecise lines are visible that only vaguely follow the shapes of the figures.

It is particularly noteworthy that the drawings were apparently executed in the already folded quires: the tip of the blade of the sword of the right fighter on fol. 3r is visible on the margin of fol. 4r.

Illustrations

199 pen-and-ink drawings in watercolour and opaque colours,[9] by four artists according to Leng; draughtsman I: 2r–42r and 58ar/v

9 Not 201 as claimed by Leng.

executed as a heretic. – Leonardo da Vinci completes *The Last Supper* on the refectory wall of Santa Mar

(draughtsman corresponds to Augsburg, JWA1, 2r–41r), draughtsman II: 52r–58v, draughtsman III: 59r–101v (corresponds to Augsburg, JWA2, 1r–42v),[10] draughtsman IV: 102r.

Although the individual sections do differ slightly in style and thoroughness of execution, it is not unlikely that the entirety of the pictures were created by only one artist. The drawing style appears to be almost identical throughout, which is particularly noticeable in the contours of the hilly backgrounds and the way the figures, swords, harnesses and crossguards are drawn.

The harnesses in both armoured combat sections are quite similar in their artistic execution, although the styles of armour differ considerably. Both sections seem to echo older models from the early decades of the 15th c., with the first one portraying a much larger variety of helmets and the harnesses showing even more textile elements from earlier eras than those in the second section. The poleyns in the first section also indicate an earlier date. Interestingly, one of these poleyns is drawn in a completely impractical and improbable way on the right leg of the left fighter on 52ra: although the knee bend is facing the beholder, the poleyn is shown from the front.

In the mounted combat section, the drawings become even rougher. To some extent, this is likely because the degree of detail needs to decrease when not only two fighters are depicted, but also their horses. Ultimately, while the artist finds their way back to a more detailed elaboration on the final page of the manuscript, the overall impression of the drawing style remains unaltered.

In summary, the artist has applied the same coarse, tentative and often hesitant strokes throughout the manuscript. Typically, they outline the figures without errors, with only a few exceptions where a number of correcting strokes or *pentimenti* were applied. The draughtsman was certainly not one of the first class; too many anatomical

10 The style of the artwork of this section does not correspond to JWA3, 3r–26r, contrary to Leng's assertion.

extravaganzas such as impossibly prolonged limbs speak clearly to this.

Iconography

All illustrations show pairs of fighters placed on a hilly lawn without shadows or backgrounds. Folios 2r–42r portray unarmoured fencers in bulbous and extraordinarily colourful robes with numerous different patterns, decorations and appliques; folios 52r–58v have rather coarse drawings of fighters in armour with hatched colouring, wearing nearly opaque tunics over their armour; folio 58a contains three uncaptioned images featuring bearded men fighting with clubs and shields, a judicial (?) duel between a man and woman and two half-naked knife fighters; 60v–84r depict roughly sketched armoured fighters with pale washes of colour; folios 85r–101v finally present similarly roughly sketched equestrian fighters, also only faintly washed with colour, with stronger colouring on the bridles and horse blankets; only the supplemental fol. 102r is more vividly coloured.

In the first part of the manuscript, the artist has taken great pains to individualise the different fencers; the only attempt made to distinguish between the armoured fighters in the two following sections was by colouring the tunics differently. In the mounted combat section, only the bridles and saddlecloths of the horses serve to distinguish between the combatants.

I strongly disagree with Leng's assertion that the images in the armoured and mounted combat section fol. 52r–102r are 'clearly borrowed from *Gladiatoria*, particularly recognizable in the dagged tunics 52r–58av'.[11] The *Gladiatoria* group consists of a number of manuscripts that deal almost exclusively with armoured combat.[12] In

11 LENG 2008, p. 87. In fact, the armoured combat ends on fol. 58v; fol. 58a shows bearded fighters, a struggle between man and woman as well as knife fighters; see above.

12 For a thorough examination of the *Gladiatoria* group see HILS 1987 and WALCZAK 2015.

Charles V, Holy Roman Emperor, is born. 1506: The antique statue of *Laocoon and His Sons* is discovered i

fact, the dagged fabrics are the only resemblance to some of the *Gladiatoria* manuscripts. Other than this, there is no similarity whatsoever, particularly since *Gladiatoria* does not contain any instructions for fighting on horseback. Any likeness is purely coincidental and should not come as a surprise when one takes the general appearance of armour into account; after all, dagged fabric parts were by no means a unique feature. Nevertheless, it does in fact come as a surprise that dagged tunics appear in Wilhalm's manuscripts. This kind of ornamental feature had gone out of fashion roughly two generations before the books were created and thus is something of a throwback to times long gone.

Leng also believes the single image of a duel between man and woman (fol. 58ar) to be a corruption of an original by Hans Talhoffer – an assumption that must also be rejected. There may very well have been different regulations about how men and women were to fight each other in a judicial setting. In Talhoffer's books, the man stands up to the waist in a pit, a club at his disposal, while the woman walks freely and has a stone in a veil which she may use to hit her opponent. The same is true for manuscript P, one of Wilhalm's probable templates. On the other hand, manuscript O, another of Wilhalm's presumed sources, has the man sitting not in a pit but in a barrel, and this is the iconography that can be found throughout Wilhalm's works.

Marginal drawings and other notes in the text

Several small drawings and notes are interspersed in the text:

1r: Note in pencil: 'cf. Cgm. 3712 f 65.'

12v: Next to 'Der Alber', a note in pencil: 'Frank'.

20v: Beneath 'Das ist der schilcher' an almost illegible name (?) written in pencil (19th/early 20th c. script): 'Breunfeid' (?).

33v: Cross potent or cross moline in the lower right corner of the page.

51r: Pen-and-ink drawing of the guild coat of arms of the Augsburg painters between the capital letters of the monogram I R; also

ome. – Works start on Donate Bramante's design for St. Peter's Basilica in Rome. – Leonardo da Vinci finishes a depiction of a trowel and hammer; underneath a rectangular cartouche with Jörg Wilhalm's closing words. Below this, there is a drawing of a fur hat, the symbol of a hatter.

58ar: An obscene drawing in very light ink in the top right corner of the lower illustration.

58bv: Several illegible scribbles, possibly the word 'Im' written down repeatedly.

59r: The year 1522 is written between the initials I and P. There's also a note in pencil: 'cf. Cgm. 3712 f 156.'

96v: Drawing of a little furry animal (a mouse or pig?).

101v: Drawing of a dragon.

Inside of the back cover: In a handwriting typical of the 19th c. (?): '102 Blätter.' ('102 sheets.')

MAD AS A HATTER – WHO WAS JÖRG WILHALM?

The most maddening thing about Jörg Wilhalm is that we know next to nothing about him – and we cannot even be certain whether the name refers to a single individual or to more than one person, possibly a father and son. Someone of that name lived in Augsburg and is mentioned a number of times in the town records between 1501 and 1575. In 1501, 1504, and 1516, the name Jörg Wilhalm appears along with his profession as a hatter.[13] Several craftsmen's files from the city archives also mention the name and profession repeatedly, at dates between 1550 and 1564. The Jörg Wilhalm recognised as a craftsman in 1501 was presumably an adult, so it is unlikely that the person recorded in 1575 or even 1564 was the same man, much older, rather than perhaps a family member of a different generation.

13 HILS 1985, pp. 191–195.

the *Mona Lisa*. 1508: Maximilian I takes the title *Elected Roman Emperor*. – Michelangelo begins painting th

There are modern-day references to *Jörg Wilhalm Huter*, in both online media and printed works, but these are incorrect. His (or their) given name was Jörg, his family name was Wilhalm, and his profession was that of a hatter – or *Huter* in German. *Huter* was not his last name.

In the years 1522 and 1523, he commissioned four illustrated manuscripts about duelling techniques in and out of armour, possibly even a fifth one considerably later, in 1556.

Wilhalm's books were created in the Renaissance, an epoch with start and end dates that cannot be clearly defined. After beginning in Italy at the end of the 14th century, it was many years before the ideas of the Renaissance found a firm grounding north of the Alps, beginning roughly one hundred years later. The Renaissance as a term from art history or philosophy overlaps with the Reformation, especially in Germany and other North-European countries.

The Renaissance brought forth such celebrities as Dante Alighieri, Niccolò Machiavelli, Martin Luther and William Shakespeare; Albrecht Dürer, Lucas Cranach, Leonardo da Vinci and Michelangelo. While it is common knowledge that Dürer and Cranach contributed to artwork of fight books, other notable but lesser-known artists such as Gregor Erhart or Georg Lemberger did so too.

Jörg Wilhalm's œuvre marks an important turning point in martial arts, duels and combat in general, thus serving as a bridge between two periods. Although created in the era of the Renaissance, it still adheres to the medieval idea of delivering old and thus venerable material to posterity. Ultimately, it exists in a realm between preservation and development.

SOURCES USED

Cgm 3711 is fed by several sources, which in turn are based on older manuscripts that may be interrelated with each other. Furthermore,

…iling of the Sistine Chapel in Rome. 1512: Martin Luther becomes a doctor of theology (*Doctor in Biblia*).

our codex has influenced later ones, which may also contain material from earlier texts. This has created a virtually inextricable web of connections, interdependencies and relationships.

For this edition, I have examined many German manuscripts from around the world – mostly in their digitised form. Since it is always a bit cumbersome to refer to them by their shelf mark, I have made use of a number of sigla, or scribal abbreviations. These are taken from an upcoming work about the entirety of Master Johannes Liechtenauer's verses, a collaboration with my colleague Cornelius Berthold. The sigla are mostly based on colloquial names, so GE stands for Gregor Erhart, or JWM2 for the second Jörg Wilhalm manuscript from Munich.

Leithandschrift – the primary source used for the comparison

JWM1 Munich, Bayerische Staatsbibliothek, Cgm 3711. [Jörg Wilhalm, 1st MS from Munich]

Manuscripts used for the concordance

JWA1 Augsburg, Universitätsbibliothek, Cod.I.6.2°.2 [Jörg Wilhalm, 1st MS from Augsburg].

JWA2 Augsburg, Universitätsbibliothek, Cod.I.6.2°.3 [Jörg Wilhalm, 2nd MS from Augsburg].

JWA3 Augsburg, Universitätsbibliothek, Cod.I.6.4°.5 [Jörg Wilhalm, 3rd MS from Augsburg].

PMD Dresden, Sächsische Landesbibliothek, Mscr. Dresd. C.93/94 [Paulus Hector Mair, Dresden].

GE Glasgow, R. L. Scott Collection, E.1939.65.354 [Gregor Erhart].

PMM Munich, Bayerische Staatsbibliothek, Cod. icon. 393 [Paulus Hector Mair, Munich].

JWM2 Munich, Bayerische Staatsbibliothek, Cgm 3712 [Jörg Wilhalm, 2nd MS from Munich].

P Paris, Musée National du Moyen Age, CL23842 [Paris MS].

PMW Vienna, Österreichische Nationalbibliothek, Cod. Vindob. 10825/6 [Paulus Hector Mair, Vienna].
JWW Wolfenbüttel, Herzog August-Bibliothek, Cod. Guelf. 38.21 Aug. 2° [Jörg Wilhalm, Wolfenbüttel].
JSW Vienna, Kunsthistorisches Museum, KK 5247 [Jeremias Schemel, Vienna].
JSWO1 Wolfenbüttel, Herzog August-Bibliothek, Cod. Guelf. 1.6.3 Aug. 2° [Jeremias Schemel, 1st MS from Wolfenbüttel].
JSWO2 Wolfenbüttel, Herzog August-Bibliothek, Cod. Guelf. 5 Blank. [Jeremias Schemel, 2nd MS from Wolfenbüttel].
O Private property; formerly kept at Tambach, Gräflich Ortenburg'sche Bibliothek, Cart. Misc. S XVI [Ortenburg MS].

TIMELINE OF THE MANUSCRIPTS

Below, I have arranged the examined sources in an approximate but probable chronological order. Some manuscripts are dated internally, while others are not. Since several codices are compendia of various independent works bound together, a number of dates can be found in one and the same volume. Even our *Leithandschrift* JWM1 consists of two parts, one from 1522, the other from 1523. Apart from that, later owners, collectors and compilers have occasionally left their own ex-libris. For example, Paulus Hector Mair, council servant and city treasurer in Augsburg in the 16th century, was an avid collector of books, especially fight books, and he was eager to leave his name in his acquisitions.

1 P: *ca.* 1480–1500. A compendium of various fighting disciplines, with very little text. Most of the illustrations have received no captions whatsoever, and those that have, were given only very brief comments. The dating by Hans-Peter Hils is based on watermarks that belong to the last two decades of the 15th century.[14]

14 Hils 1985, p. 47.

s theory of heliocentrism. 1516: Thomas More publishes *Utopia*. 1517: Martin Luther posts his 95 Theses

2 **O**: 1515. Apart from fighting techniques in armour and on horseback resembling Wilhalm's, this manuscript contains the teachings of Johannes Liechtenauer (unarmoured, armoured and mounted combat), wrestling by Ott the Jew, Martin Hundfeld's armoured fighting on the ground and on horseback, techniques for the axe, the large shield, the sword and buckler, and the dagger.

1v: 1558: 'Joachim Graue zu Ortenburg.' ('Joachim, Count of Ortenburg.')

171r: 'den zehenden tag des monatz februarij 1515. J. [?] Schwerer.' ('10 February 1515. J. [?] Schwerer.')

3 **JWA3**: 1522. Allegedly the template for **JWA2**. The manuscript contains very little text, some of which has been cut away by cropping the pages. There are several numbers on each page, many of them crossed out, which hints at an order that was constantly under revision. Hils claims that the date is written on fol. 1r.[15] The manuscript has obviously been rebound since his inspection, and the sheet in question is no longer in its former place, but rather is now a loose supplemental sheet.

Book cover: '1552 Paulus Hector Mair.'

Fragment 1 (formerly fol. 1r): '1522.' (The date is repeated underneath the original handwriting in a more modern hand.)

4 **JWA2**: 1522. Jörg Wilhalm's more elaborate version of codex **JWA3**, which describes fighting in armour and on horseback in words and pictures.

Iv: 1561 'P. H. M.' (i. e. Paulus Hector Mair.)

1r: 'Vonn Jerg schefflers Erben vber kumben durch den lenhart Zolinger vnd Sie zu danckh bezolt am 7 december annum 1561 Jar.' ('Via Lienhart Sollinger taken over by Jörg Scheffler's heirs, who I gratefully paid on 7 December 1561.')

42v: '1522.'

15 Hils 1985, p. 37.

on the door of the Wittenberg Castle Church, thus starting the Protestant Reformation in Germany. 151

43r: '1561 Paulus hector Mair zu geherig.' ('Paulus Hector Mair's property.')

5 JWA1: 1523.

2r: '1544 Jar Das buch Ist des Jorg wilhalms huders gewest vnnd Inn dem Jars hab ichs Iber khumben p. h. mair.' ('The year 1544. The book had belonged to the hatter Jörg Wilhalm, and in that year I received it. P. H. Mair.')

41r, 44r: '1523.'

41v: '1544 paulus hector mair zugeherig Ist des huders gewest.' ('Paulus Hector Mair's property. Had belonged to the hatter.')

71r: '1564 Liennhartt Sollinger.'

6 JWM1: 1522/1523. The *Leithandschrift*, subject of this edition.

59r: '1522.'

42r, 45r, 51r: '1523.'

7 GE: 1533. A compilation of numerous fencing teachings, including those of Jörg Wilhalm and Johannes Liechtenauer, possibly authored by the sculptor Gregor Erhart whose name is referred to in the first person singular on fol 2r.

2r, 111r, 112r: '1533.'

3r: 'Im 1560 Jar am sannth michels tag hab ich paulus hector mair / das buch durch denn Lenhart Zolinger meserschmit vber khumben vnd In darumb vergniegt.' ('On Michaelmas Day in 1560 (29 September), I, Paulus Hector Mair, bought the book from the cutler Lienhard Sollinger and compensated him for it.')

189r: The manuscript contains three versions of Liechtenauer's teachings. At the end of the third, which is otherwise mostly identical to version 2 in JWM1, an author's name appears that is absent from every other source:

'Deo graczs amen Nicolaus thum im 1486 iare in vigilia maria virginis visitationis.' ('God have mercy, amen. Nicolaus Thum, the eve of the Visitation of Mary (2 July) in the year 1486.')

8 PMD: After 1542, before 1579. Compendium consisting of two volumes in the German language, commissioned by Paulus Hector Mair. Giving an exact date for this and the two following works is challenging. The artist Jörg Breu the Younger is usually believed to be the author of the illuminations in all three double-volume compendia. He died in 1547, while Mair only purchased the last of the sources that were used for the creation of his books in 1566. So the three double-volumes were either produced after that date, or the source material had been accessible to the artist – be it Breu or somebody else – before.[16] Since Mair was executed in 1579 because he had helped himself too freely to the city treasury of Augsburg, the books must have been created before this.

C.94, 196r: 'Wie aber die bemellte Sannct Leonharts kirch Anno 1542 von dem Rathe zu Augsburg abgebrochen …' ('As the afore mentioned church of St. Leonard has been dismantled by order of the Augsburg city council in 1542 …')

9 PMM: After 1542, before 1579. Two volumes in Latin language. The most extensive and elaborate compilation of Mair's.

Volume II, 162v: '… at cum ex mandato Senatus Augustani anno 1542 id templum [d. Leonardi] destrueretur …' (Translation: see above, No. 8.)

10 PMW: After 1542, before 1579. Mair's bilingual Latin–German version in two volumes.

10826, 195r: 'Wie aber die bemelte Sanct Leonharts kirch Anno 1542 von dem Rate zu Augspurg abgebrochen …' (Translation: see above, No. 8.)

203r: '… at cum ex mandato Senatus Augustani Anno 1542 id templum [D. Leonardi] destrueretur …' (Translation: see above, No. 8.)

16 For more information on Paulus Hector Mair and his manuscripts, see HAGEDORN/AMBERGER 2020.

excommunicates Martin Luther. Luther then hides at Wartburg castle, where he begins his translation of t

11 JWM2: 1556. A collated manuscript, formed out of various codicological entities: the first part (1r–52v) is based on the fight book attributed to Lew the Jew;[17] the second contains Johannes Leckúchner's teachings for the *Messer*,[18] although they are attributed to Master Liechtenauer, and several other sections (some of them out of sequence) about the sword, dagger and wrestling that are in one way or the other related to the Masters Liechtenauer or Ott (53r–88v); the third (89r–96v) is dated to 1556 on fol. 95v and contains two more versions of Liechtenauer's long sword techniques (verses only as well as combined verses and glosses) which are next to identical to the versions in our manuscript JWM1; the fourth mirrors the entire illustrated material from JWA1, although it is only drawn in outlines (97r–195v); the fifth shows uncaptioned sword and buckler plays (196r–212r, except 210r–211v); a brief sixth interrupts the previous one and offers a few uncaptioned images of Wilhalm's miscellanea (210r–211v). Various illustrators and scribes make a clear distinction of these units difficult. Due to the variety of the individual parts, a definite dating is impossible to determine, except for unit 3.

25r, 48r, 101r, 109r, 123v, 166v: 'Johan Jacob Spaney [Straubing] Camerdiener S.E. 1730' (chamberlain of his excellency) – from the hand of a later owner.

95v: 1556 (end of codicological unit 3).

201r: 'Johann Ehrnßhaimer geborner zu Ehretshaim Obrister Wachtmeister Comentiret alhir zu Straubing den 2. Juni 1634' – from the hand of a later owner. ('Johann Ehrnßhaimer, born in Ehretshaim; Supreme Constable, comments here in Straubing, 2 June 1634.')

12 JWW: First half of the 16th century/around 1588.

150r: The fight book is supplemented by a family history of the

17 For a detailed analysis, see the author's edition of the manuscript: Hagedorn 2017.

18 For a definition, see the glossary.

ble. 1522: The Knights' Revolt erupts in Germany. – Martin Luther's translation of the New Testament into

Fuggers, a wealthy trading family from Augsburg. The last entry in the Fugger family chronicles mentions the year 1588 as the date of birth of one Jacob Fugger. But since this part is an entirely different codicological unit, the first fight book part may be either older or – although unlikely – younger. The clothing style of the unarmoured fencers points to the first half of the 16th century. The drawings are cruder than in GE but nevertheless similar: Jörg Wilhalm's teachings for the sword, in armour and on horseback are drawn in simple yet clear outlines. The codex is a compendium and also contains a fencing treatise by Lienhart Sollinger, as well as a version of Liechtenauer's verses and glosses. Curiously, it is Sollinger's name, not Wilhalm's, that appears at the end of this section.

13 JSWO1: After 1568. A massive illustrated volume that concentrates primarily on equestrian matters, including brief reports of tournaments, the first being held in the year 938 – although the Renaissance armour shown in the image is anachronistic to this early date. Other sections show garnitures of tournament armour and Wilhalm's armoured and mounted techniques. Fol. 400r mentions a tournament on the occasion of the marriage of William V, Duke of Bavaria and Renata of Lorraine in Munich (22 February 1568) which is the latest date given in the manuscript and should therefore be considered the *terminus post quem*.

14 JSW: The data provided by the Kunsthistorisches Museum in Vienna suggest that the content of the manuscript is more or less identical, although slightly less extensive. Therefore a similar date of creation can be assumed.

15 JSWO2: The same applies to this manuscript from the Herzog August-Bibliothek in Wolfenbüttel.

The following stemma depicts the connections of the manuscripts among each other. Since no direct master/copy relationship can be reliably established, the connecting lines do not define dependencies as in a family tree, but rather indicate contextual similarities.

Early New High German is published. 1523: Gustav Vasa is elected king of Sweden, thus establishing t

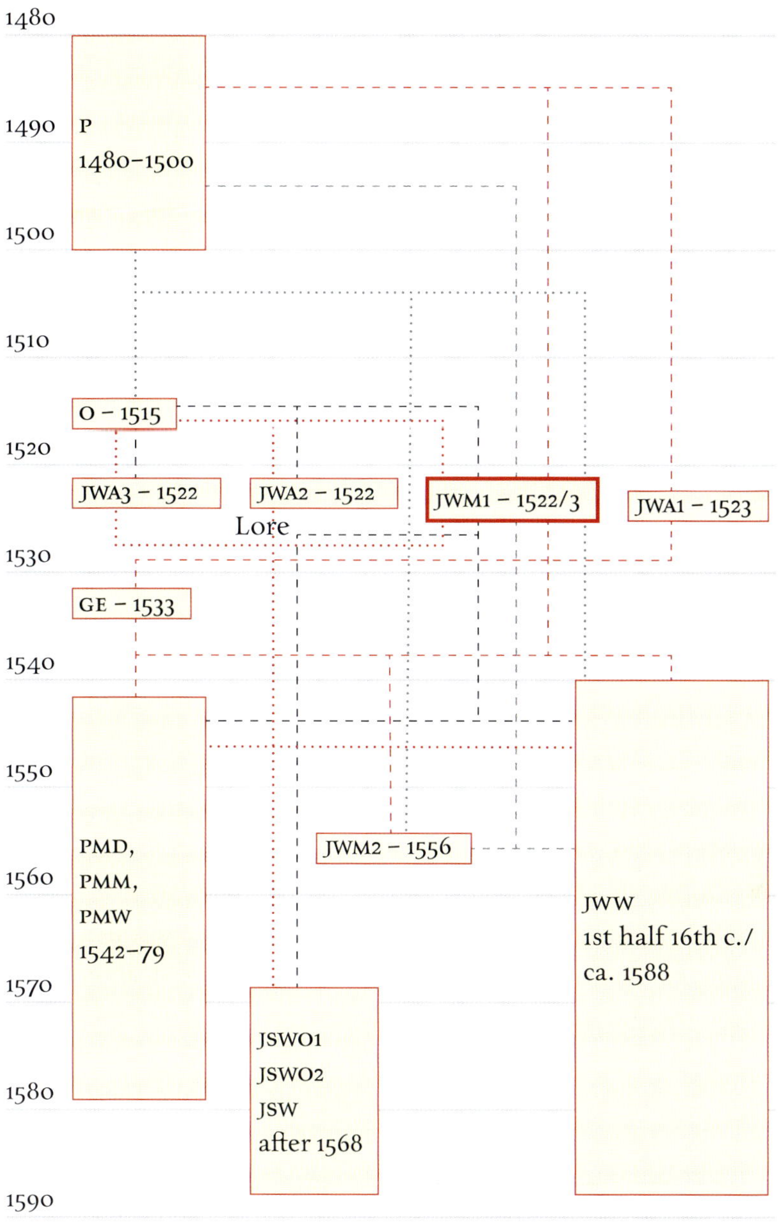

Sword: - - - Harness I: - - - Harness II: - - - Miscellaneous: · · · · Mounted combat: · · · ·

lependence of Sweden from Denmark. 1524: The beginning of the German Peasants' War, a widespread

ANGELS, DEVILS AND PRETZEL PEOPLE – WILHALM'S HUMAN MENAGERIE

It is remarkable that the Augsburg codex JWA1 features the same unarmoured sword fighting techniques as the *Leithandschrift*, although, due to a faulty rebinding of the manuscript, these are frequently out of sequence. The text is next to identical, but the images differ vastly. It is not the postures that are so dramatically different, but the costumes. While JWA1 itself features an enormous array of flamboyant and eccentric clothing, this is by far surpassed by JWM1, our manuscript. The colours in the Augsburg manuscript are mostly washes of bright red, rose, black, light grey and light yellow with very rare instances of powder blue. In contrast, JWM1 offers a much wider palette of generally more vibrant colours, among them not only bright reds, yellows, blues and greens, but also strong blacks and greys; paler hues of rose, grey and yellow appear as well. It's the huge and frequently bizarre variety of costumes that makes our manuscript truly stand out. No other known exemplar offers this much individual detail, and goes beyond the simple depiction of fashion. We encounter all kinds of grotesque figures, and even angels and devils, throughout the pages. Particularly noteworthy are:

- Fighters with (deliberately?) ragged or torn clothing – imitating the dashing and daredevil *Landsknecht* style, with occasional bare legs or feet: 2v, 4v, 6r, 7v, 8r (with a bleeding wound on the leg), 8v, 10r, 12r (another leg wound), 16v, 18r, 21r, 22r, 22v, 24r, 26r, 27r, 27v, 28r, 30v, 32r, 33r, 33v, 34r, 35r, 37v, 38v, 41v.
- Bizarre beards: 2v, 3r, 10v, 21r, 24r, 29r, 41r.
- Emblems, symbols and animals as decoration on the garments:
  - Winged beast: 3r.
  - Owl and three black birds: 5r.
  - Blue bird (eagle?): 6r.
  - Text scroll: 8v, 42r.
  - Tree: 10r.

revolt in several German-speaking areas across Central Europe. 1525: The last Aztec Emperor, Cuauhtém

    - Monkey on a red ball: 13r.
    - Sun and flames: 13v.
    - A face: 22v.
    - Lobsters: 24v.
    - Snake on a cushion, with the letters Q and S next to it: 27r.
    - Iron (?) grate: 29r.
    - Flowery, almost jellyfish-shaped patterns: 35r.
- Expansive hats: 3v, 13v, 31r, 40r.
- Bells – though it is unclear whether they are real ones or patterns on the fabric: 4v, 6r, 7v, 10r, 11r, 13v, 15v, 26r.
- Person in a demi-long habit with patches and a badge, wearing a dunce's cap and one torn shoe: 7v.
- Fighter (possibly *alla turca*) with a turban and a sabre or falchion with a pommel in the shape of the head of a predatory animal or a dragon: 10v.
- A dwarf or child next to the two fighters without any significance for the depicted fighting technique. This figure is absent in all comparable sources. The barefooted person is in a dress bordered with black wavy patterns and holds two objects in the hands. The one in the left hand could be a bag or a purse. The one in the right is a bit more mysterious: due to its entwined appearance it has been identified as a pretzel and the person holding it as a baker or pretzel seller. In fact, it resembles the guild symbol of the Augsburg bakers: 11r.[19]
- An angel, complete with wings: 12r.
- Lose scabbard on the ground: 16v.
- Short curved knife on a low hanging belt: 18r.
- An open book on the ground: 19v.
- Person wearing clothes from eastern Europe: 27r.
- Fighter wearing an iron hat, breastplate with long tassets and pattens: 29r.

19 I'm obliged to Christian Trosclair for pointing this out to me.

killed by Hernán Cortés. 1525: The German Peasants' War ends with the defeat of the peasants, over

- Person with coat and chaperon: 34r.
- Fool's cap – both as an ornament on the hose as well as the real thing, hanging over the shoulders at the back: 36r.
- The devil – with horns, sharp teeth, wings and claws, falling souls on the tunic: 37r.
- Person in black and white garments (probably not a cleric) with bare calves and a fur hat (?) between his legs: 40v.

While the artist is the same person who produced the not-quite-so exuberant illustrations in JWA1, it remains entirely unclear why a different approach was chosen in the creation of the two manuscripts – particularly considering that the images of armoured and mounted combat are interchangeably similar in the two sister manuscripts JWM1 and JWA2.

JOHANNES LIECHTENAUER – AGAIN AND AGAIN

Johannes Liechtenauer is *the* authority as far as the art of fighting in medieval Germany is concerned. His teachings survive in 28 out of approximately 80 manuscripts or early printed sources from the 14th to the 16th century and are recorded in the form of rhymed couplets, known as *Zedel* – a word that derives from Latin *schedula* and signifies a more ephemeral form of notation, than for instance, a book. A German *Zettel* nowadays would be a note, nothing more than a slip of paper.

It is a peculiarity of Liechtenauer's Notes that they come in two versions: one consists solely of verses that may have been put down by Liechtenauer himself in the first place. The second version keeps the verses but adds extensive commentaries, referred to as glosses. These are most likely not authored by Liechtenauer himself. Both versions exist independently. Some sources only render the verses, others include the glosses as well, and a few include both versions;

75,000 of whom have lost their lives. 1527: The High Renaissance ends with the Sack of Rome, by Spanish a

usually the verses-only version comes first, being followed by the commentated one.

Since his work was so popular, it seems only natural that Liechtenauer should have received a proper pictorial treatment. In fact, there are a number of sources that add images to his teachings. Fencing masters like Paulus Kal or Hans Talhoffer created extensively illustrated manuscripts, but they only contain brief captions, in most cases only loosely based on Liechtenauer, that hardly serve to explain the action displayed in the image. Other masters, such as Peter Falkner, came up with a different idea. His manuscript (PF) offers a mixture of Liechtenauer's verses and those by Johannes Leckuchner, a fencing master who exclusively taught *Messer* fencing. Leckuchner relies heavily on the pattern laid out by Liechtenauer and his glossators: brief verses are followed by extensive commentaries. His first manuscript from 1478 (**LH**) offers only text, while the second one from 1482 (**LM**) accompanies each technique with an illustration. Leckuchner, however, did not invent this arrangement. There are a number of additional manuscripts that follow the same strategy, such as the very first surviving fight book from the early 14th century, I, as well as the three *Gladiatoria* manuscripts **GK**, **GN** and **GW**, or Baumann's fight book (B). Wilhalm has developed something quite similar: he took up Liechtenauer's verses on the long sword, and to some extent the glosses too, broke them down into rather small consumable portions and added an individual illustration to every single technique. Our manuscript **JWM1** uses Johannes Liechtenauer's *Zedel* as a template in the following sections: foreword (1r), illustrated sword fencing (2r–42r), verses (43r–45r) and verses and glosses (45v–51r). In this way, Wilhalm preserved Liechtenauer's art as it was known, and as a bonus, he created elaborate picture books that by far surpassed their only known predecessor: manuscript SE from Glasgow has a considerable number of illustrations too, but these are interspersed into the original and unaltered glosses of Master Liechtenauer's teachings.

erman troops. Pope Clement VII is forced to make peace with Charles V. 1529: Vienna is besieged by the

To elaborate in detail on the family tree of texts that rely on Master Liechtenauer is beyond the scope of this introduction, but will be the subject of an upcoming publication. Nevertheless, I have already given a brief outline in my edition of the Lew the Jew fight book[20] in which I have identified three groups of Liechtenauer-related sources, with some additional independent individual works. This investigation, however, was based exclusively on sources that contain both Liechtenauer's verses *and* the glosses. Unfortunately, I had overlooked two sources, and another one came to my attention only recently. Taking these additional sources into consideration, along with the verses-only versions, we arrive at seven distinguishable lines of tradition. Group I, the tradition line of Lew the Jew: 8 manuscripts; group II, Peter von Danzig: 2 manuscripts; group III, Master Sigmund: 4 manuscripts; group IV, Paulus Kal: 4 manuscripts; group V, Nicolaus/Wilhalm: 5 manuscripts; group VI, Jobst von Württemberg: 2 manuscripts; group VII, Andre Paurenfeindt: 2 prints, 1 manuscript; independent or not clearly classifiable sources: 5 manuscripts.

The Notes are delivered in three disciplines: the first is about unarmoured long sword fencing, the second explains mounted combat, and the third outlines how to fight in armour.

Jörg Wilhalm only makes use of Liechtenauer's teachings for the long sword; sections dealing with armoured and mounted combat go back to another source, probably codex **O**.

PREDECESSORS AND SUCCESSORS

Jörg Wilhalm's œuvre does not exist in a vacuum; he relies on older sources, and his successors in return rely on him. Wilhalm's books are relatively young when compared to the oldest extant fight book, a manuscript about fighting with sword and buckler from the

20 HAGEDORN 2017.

beginning of the 14th century which is – unlike any other fight book – almost exclusively referred to by its shelf number I.33 (abbreviation I).[21] Roughly two hundred years separated Jörg Wilhalm from the anonymous author of I, and a lot of fencing material had been written in the meantime. A common approach was to copy from older sources, sometimes unaltered, sometimes enhanced and adapted to the fashionable fencing style of the new time. Also, the appropriation of older sources was a frequent procedure among medieval fight book authors or compilers. This was done in two contrasting ways: either the authors referred to a revered and famous old master, such as Johannes Liechtenauer, by name in order to add more credibility to their teaching and to back it up with an unquestionable authority; or they hijacked and plagiarised other people's teachings and passed them off as their own invention. In the end, it becomes increasingly difficult to disentangle the resulting mess. In some instances, it remains impossible to determine the original and true author, meaning that we are left with speculation.

The overwhelming majority of German sources unequivocally attributes the famous verses about fighting techniques to Master Johannes Liechtenauer, such as the Peter von Danzig manuscript (PD, dated to 1452) on fol. 3r:

> 'Alhye hebt sich an dye zedel der Ritterlichen kunst des fechtens dye do geticht vnd gemacht hat Johans Liechtenawer der ain hocher maister In den künsten gewesen ist dem got genadig sey.'
>
> 'Here begin the Notes of the chivalric art of fencing that have been versified and created by Johannes Liechtenauer who was a great master of these arts; may God have mercy on him.'

21 Cinato 2016, Forgeng 2018.

cuments of the Protestant Reformation, is presented to Charles V. 1533: Ivan IV (the Terrible) becomes

On fol. 2r of one of his manuscripts (HTK from 1459), fencing master Hans Talhoffer presents the verses with only minor changes, which nevertheless led him to attribute them to himself:

> 'Hie lert der talhofer ain gemaine ler in dem langen Schwert von der zetel.'

> 'Here Talhoffer teaches a common lesson of the long sword, taken from the Notes.'

Due to the ambiguous structure of the sentence, it may be that Talhoffer only sees himself as a teacher who makes use of the Notes without explicitly claiming authorship; nevertheless, he entirely neglects to mention Liechtenauer's name.

Another account can be found in a manuscript from Jena (J, dated to 1418–28). On fol. 123v, it says:

> 'Sequitur bona et vera moda dimicandi magistri · h · Beringois pie memorie.'

> 'Here follow the good and true ways of fencing by Master H. Beringer, held in dear memory.'

The rendition of the Notes in this source is peculiar not only because it credits a certain Master Beringer we have no other knowledge about, but also because the rhyming couplets are presented in a completely different order; a few are missing, and some new ones have been added.

Jörg Wilhalm also copied from older fight books, and he too posed as the alleged author. At the end of an altered but still recognizable version of Liechtenauer's foreword, he names himself as the author (fol. 1r). But then again, he also gives credit where credit is due. In our manuscript, for instance, on fol. 45r and 45v, he twice mentions

Grand Prince of Moscow – and later the first Tsar of Russia. 1534: Break with Rome and establishment of t

Johannes Liechtenauer as the author of the teachings on the long sword which Wilhalm had written down in two versions, one containing only Liechtenauer's verses, the other a version with additional explanatory glosses.

The composition of Wilhalm's book can be traced to a number of sources. The two large illustrated parts of fighting in armour and on horseback appear a couple of years earlier, in the so-called Ortenburg manuscript (O, dated to 1515). Wilhalm copies the images almost without alteration despite one curious exception: in the armoured combat section, the two opponents are bordered by wooden lists on the left and the right, indicating the limits within which a judicial duel was being executed. Wilhalm copies the text, but again he omits a brief introduction to the armoured combat on fol 86r:

> 'Item das ist der ernst vnd trutzlichist kampff / So vor zeittem viel geschehen ist vmb Eren willenn / oder welcher sich in solchen trutz aber kampff begeben hatt / ist hinfuren zw keinen Ritterlichen vnd erlichen sachenn gebraucht worden / aus vrsach die ich nit vermain nott Sein zw endecken / sunder las ich es die weissen vnd die meist*er* des kampffs aus Richtenn.'

> 'This is about the serious and most bold duel, of which there were many in the old times and in which honour was at stake. But whoever proudly renounced such a fight was thereafter unable to enjoy chivalrous and honour-related matters, for reasons which I do not consider necessary to elaborate, but let the scholars and the masters of the fight explain them.'

So even manuscript O in 1515 considers the depicted techniques to be old already. Nonetheless, they were copied time and again. Wilhalm in 1522/23 was by no means the last. Another preserver of the old arts was the afore-mentioned Paulus Hector Mair. He collected fight books that were up to a century old and recreated the techniques

hurch of England during the reign of King Henry VIII. 1536: The Inquisition is implemented in Portugal.

therein in his own fashion. He did so by hiring skilled swordsmen who were to pose for gifted artists, who in return created some of the most elaborate artwork concerning fencing matters that have survived up to this day. It was Mair's desire to preserve the old arts and bestow them on posterity so that particularly the youth would learn them and thus lead virtuous and brave lives.[22] Mair's last known acquisition was marked with ownership by his own hand in 1566 (manuscript SS). He obviously saw the centuries-old sources as fascinating relics of the past, worthy of transmission, even worthy of recreation. Modern practitioners of the historical European martial arts such as HEMA certainly have much in common with Mair.

Manuscript P from Paris (*ca.* 1480–1500) is another one of Wilhalm's predecessors. Although the illustrations seem to be a refined and more accurate version of those in Peter Falkner's manuscript (PF, *ca.* 1480–1500), the long sword, *Messer* and dagger sections only contain coincidental similarities. Interestingly, matters are different with other parts, such as the mounted combat section: although P is more extensive and one of the combatants is displayed without armour, there is an undeniable similarity and even nearly identical portrayal of several techniques: nine out of ten pictures in PF can also be found in P, which contains a total of 25 mounted techniques. Another section of only four images about duelling with large spiky shields is absolutely identical in both P and PF; three out of six halberd techniques from PF occur in P, which contains only four techniques in all.

In summary, P and PF share some common origin; P has given more than half of its unarmoured sword fighting techniques to Wilhalm; a relationship of the armoured and mounted combat sections cannot be satisfyingly determined, since occasional similarities occur but may describe generic postures that are inherent to fighting in armour or on horseback. In this regard, Hils is mistaken when he claims that the section about harness fencing must be identified as

22 See for instance PMD, fol. 2r.

originating from P.[23] Nevertheless, two images at least, one showing half-naked men with short knives, the other a duel between man and woman, have their identical counterparts in Wilhalm's works.

While Wilhalm compiled his volumes from various sources, other authors did the same and incorporated Wilhalm's material into their works. One of the more prominent ones may have been Gregor Erhart (*ca.* 1468–1540), a famous German sculptor from Ulm, who features the same unarmoured long sword section, but adds a lot of extra material to his extensive collection of texts. It is notable that although Erhart's artwork consists only of quickly drawn outlines, these are nevertheless artistically superior to the carefully coloured and washed pictures in Wilhalm's works. The possibility remains that both Wilhalm and Erhart independently copied from a common predecessor. Furthermore, Paulus Hector Mair incorporated material from Wilhalm into his magnificent volumes, as did Jeremias Schemel around the middle of the second half of the 16th century.

The Wilhalm group: differences and similarities

Hans-Peter Hils suggests that the first part (fol. 1–58a) of JWM1 is an archetypal version, created in 1523, which is also contained in JWA1. He claims that the second part (fol. 59–101) is a copy of either JWA2 or JWA3.[24] This, however, can neither be affirmed nor denied with certainty.

As we have seen, the watermarks in JWM1 do not match their respective counterparts in JWA1 and JWA2. But since JWM1 also sports the ox head watermark of the two sister manuscripts numerous times, it can be presumed that the artist and the scribe had access to a variety of papers in their workshop.

JWA1 and the first part of JWM1 are extremely similar; the artist and the scribe are the same in both manuscripts. The only major

23 Hils 1985, p. 194.
24 Hils 1985, p. 99.

differences are the above-mentioned more flamboyant and fanciful costumes in our *Leithandschrift* and the accuracy of the handwriting, which is a bit more precise and clear in JWA1. There are some omissions in the text in JWM1 in comparison to JWA1, although the same is also true the other way round, but to a lesser extent.

A small hint that JWA1 comes before JWM1 can be found in the repeated occurence of the phrase 'as it is painted below' in JWA1 (20v, 27r, 29r, 34r, 37r) despite the image referred to being in fact *above* the text. The scribe has possibly copied from a source similar to P – which has hardly any text, but plenty of space left for it above the images. Noticing his faux pas, the scribe corrected 'above' to 'below' in the subsequent manuscript JWM1.

When we compare the second part of JWM1 to JWA2, the drawing and writing styles are nearly identical once again, and it is almost impossible to determine which manuscript came first. The postures of the armoured figures and the colour of their tunics do not differ at all. One small difference is that in the mounted section, the illustrations in JWA2 seem to have been executed with a trifle more care; furthermore, they are numbered within the image. However, it cannot be said with certainty that these numbers come from the same person who wrote the rest of the text: they are less slanted and slimmer and appear to be penned in a slightly darker ink.

In any case, it appears as if both parties, illustrator and scribe, worked on the two manuscripts JWM1 and JWA2 more or less simultaneously. While we cannot determine whether either served as the template for the other – if at all – it is not unlikely that both are based on JWA3, the so-called sketchbook version with its crossed-out images and numbers as well as only rudimentary texts, if any. It is interesting, however, that the artwork in JWA3 is much more refined than in the assumed final works, which both show much less artistic skill. Another intriguing detail are the rondels on the sword blades in JWA3, which exist neither in JWA2 nor in JWM1 – but do appear in O. These objects possibly support a firm grip of the left hand on the

of the Schmalkaldic League. – Michelangelo becomes the chief architect of St. Peter's Basilica in Ro

blade and might even offer some hand protection. Thus, the swords are clearly identified as specific weapons used in a duelling situation in which the half-sword grip was employed. The artwork in O and JWA3 is much more similar to each other than to the two later versions from 1522, JWA2 and JWM1.

One curious detail is folio 28 in JWA3, which differs from the rest of the manuscript in that it shows only outline drawings on the recto and the verso, while the rest is carefully coloured. These two drawings reappear only in JWM2 (fol. 155r/v) and in our *Leithandschrift* as the two more brightly coloured supplemental pictures on fol. 102r.

THE TEXT

Jörg Wilhalm collated his manuscript from a number of authors, some of whom are mentioned by name, while others are not. We can also make out direct connections as well as merely indirect relationships to additional sources by carefully comparing their pictures or texts. We are left with quite a number of uncertainties and cannot answer whether any manuscript is an immediate copy of another. It is quite possible that various artists copied from one or more common sources independently. Nevertheless, it is more than likely that at least Paulus Hector Mair used JWA1, JWA2, JWA3 and GE for the creation of his own works PMD, PMM and PMW, since he kept these manuscripts in his library. While it is true that most of them only came into his possession later, this nonetheless suggests that Mair was at least familiar with them when creating his works.[25]

The text of our manuscript consists of at least two codicological units from 1522 and 1523 respectively. The later part, dated to 1523 on fols. 42r, 45r and 51r in quire 2 (letter P watermark), describes unarmoured fighting techniques, separated into an illustrated and

25 See Hagedorn/Amberger 2020, pp. 30–31.

captioned version and another one that only has text. The earlier part is dated to 1522 on folio 59r in quire 5 (pine cone watermark) and contains armoured and mounted combat. Quires 3 and 4 (the latter being entirely empty) do not carry any dates. They possibly belong to the later codicological unit due to an overlap of watermarks: four sheets of quire 2 show an ox head watermark like quires 3 and 4 – and the empty quire 1. However, this cannot be known for sure, and the existence of a third codicological unit is not impossible.

In the course of the centuries, Master Liechtenauer's verses have deteriorated considerably. Generations of scribes have continually copied from their respective sources, and perpetual misreadings, frequent misunderstandings, occasional omissions and constant attempts to correct an obvious – or assumed – corrupted master have led to a prodigious number of versions and an even more prodigious number of flaws. The tradition line of Jörg Wilhalm is among the latest ones and consequently bears the most significant and numerous artefacts.

Wilhalm has copied the Notes from sources as yet unidentified. His rendition of the verses alone doesn't offer much insight into its exact origin; it does not follow any of the other Liechtenauerian tradition lines precisely. A closer look at the glosses, however, reveals a clearer picture. About half of the text corresponds to the overarching verbalisation of the glosses which can be found in any of the sources; a small portion parallels the Peter von Danzig version (group II); an even smaller fraction is completely unique to Wilhalm or consists of obvious misreadings and corrections of garbled phrases; but roughly a third conforms to the tradition line of Lew the Jew (group I). We can therefore assume that the most widespread group has spawned another with enough individuality to stand on its own.

The organisation of the glosses and the order of the techniques contained in them, however, differ significantly from all other traditions: they start with the final section, resume the text in the middle, and the beginning is completely absent, so that the *Schieler* is the first

first almanac. 1555: At the Diet of Augsburg (the meeting of the Imperial Diet of the Holy Roman Empir

technique described, whereas in other sources, the *Schieler* is preceded by an introduction and extensive descriptions of three other technique complexes, namely the *Zornhaw*, *Krumphaw* and *Zwerchhaw*, which are, curiously, nonetheless contained in the illustrated section.

The following overview shall clarify which text version is contained in which manuscript within the Wilhalm group (for the images, see the concordance, p. 394).

Foreword to the illustrated long sword

JWM1: 1r

GE: 2r

Liechtenauer's verses

JWM1: 43r–45r

JWA1: 42r–44r

GE: 167r–171r (with preface)

JWM2: 65r–66r, 78v–79v (with preface) – The text in this manuscript is occasionally substantially out of sequence.

JWM2: 89r–91r

JWW: 48r–49v

Liechtenauer's verses and glosses

JWM1: 45v–51r, attributed to Jörg Wilhalm

JWA1: 44v–49r, attributed to Jörg Wilhalm

GE: 179r–189r, attributed to Nicolaus Thum

JWM2: 91r–95v, attributed to Jorg [!] Wilhalm

JWW: 50r–53v, attributed to Lienhart Sollinger

When we look at the manuscripts of the Nicolaus/Wilhalm tradition,[26] it is Gregor Erhart's version (GE, from 1533) that seems to

26 Since GE ascribes the glossed version of the Notes to Nicolaus Thum, accompanied by the date 1486, it is possible that Wilhalm based his long sword material on a source that was almost 40 years old, authored by Nicolaus.

e Peace of Augsburg is signed between Charles V, Holy Roman Emperor, and the Lutheran Schmalkaldic

recite the Liechtenauerian texts in the least deformed way. **JWA1** and **JWM1** follow next, while **JWM2** and **JWW** are even further from the original. **GE** is also the only other version, next to the *Leithandschrift*, to contain the foreword (2r) for the illustrated part.

Obviously, later authors, such as Wilhalm, made good use of the handed-down material from their predecessors. Sometimes they copied wrongly; sometimes they misread the template; sometimes they made amendments that may or may not have been improvements; sometimes they tried to correct corrupt syntax and invented something new that made sense to them and that was passed on in turn; sometimes they came up with clever new solutions to old problems. In this way, particularly Wilhalm's teachings offer an astounding insight into the long-lasting fascination for personal combat techniques. Wilhalm takes up the torch and relays it in an unprecedented way. He follows his predecessors insofar as he gives us a rendition of Liechtenauer's verses which is followed by their repetition with explanatory glosses. What makes Wilhalm truly stand out is that he adds a huge number of images. Illustrated manuscripts in the tradition lines of Johannes Lichtenauer that predate Wilhalm are rather scarce. In fact, we only know of **SE**, dated to 1508. A later exemplar is the so-called Goliath manuscript **G** (1535–40) which is based in large part on **PD** (1452). These examples integrate a fair number of images into the text. Wilhalm, on the other hand, follows an entirely different approach: he offers us a third version of Liechtenauer's verses, one that picks up many of his verses, supplemented by similar or varying glosses and a vast number of pictures. In this way, Wilhalm has created something that had not existed before in the Liechtenauerian tradition. One might argue that his interpretation of the source material is not wentirely in line with earlier glosses, but then again, one must take into consideration that his corpus stems from another era. In any case, his images give valuable insight into the execution of many a technique.

League. The treaty establishes the principle of *cuius regio, eius religio*, which enabled the princes to decide t

In this way, the teachings of Master Johannes Liechtenauer went through many copyists' hands and were always inspired by new input. Their impact continues even today – in the hands of recreators of the old art, the practitioners of HEMA.

CONCLUSION

Jörg Wilhalm transfers the centuries-old art of combat in and out of armour to the Renaissance period. He creates a new method of portraying Master Johannes Liechtenauer's teachings for the long sword by picking up the texts, modifying them and accompanying them with a great many images. A similar approach had already been taken by authors of fight books following different tradition lines.

Wilhalm's work would not have been possible without its many sources, which are sometimes known but not always identifiable. Wilhalm relied on material by Johannes Liechtenauer, Nicolaus Thum or J. Schwerer, while his books served as an inspiration for compilers such as Paulus Hector Mair in return.

His works that carry dates stem from 1522/23 and 1556, but it has not been possible to determine unambiguously whether they should be seen as early and late books by one individual, or as volumes created by father and son, or whether a later copyist simply forwarded the name of the original author to posterity.

With his colourful and inventive illustrations, particularly in our manuscript, Jörg Wilhalm certainly sparks the imagination of the beholder and increases interest in the fighting techniques that, from a modern person's view, links the Middle Ages to the Renaissance.

ligion of their subjects. 1556: Charles V abdicates and is succeeded by his brother Ferdinand I.

Jörg Wilhalm

· THE ·
MANUSCRIPT

†

EDITORIAL GUIDELINES

THE TRANSCRIPTION

The transcription corresponds to the original to the letter, and line breaks are indicated by a grey pipe (|). Abbreviations are resolved and marked by *italics*, completions of missing text passages and those added from JWM1 are set in [square brackets]. Occasional special characters or diacritical markings above a letter have mostly been ignored, such as an *Umlaut* above y or a demi-circle above the letter u which was used in the manuscript to distinguish it from the otherwise identically written n. Other *Umlaute* (ä, ö, ü) have been retained, since they alter the pronunciation of the vowel. As was customary at the time, an initial vowel 'u' was written as 'v'; the transcription sticks to this original 'v'. The capitalisation reflects the original, but when in doubt, lowercase has been preferred. Capital 'I' and 'J' (the same letter in Gothic script) are used according to the conventions of modern New High German. The same procedure was chosen for capital 'U' and 'V', which are also the same letter in the original – unlike their lowercase counterparts – and which are also rendered according to modern standards. Letters, words or passages that were cancelled in the original are represented by ~~crossed out~~ text. To accommodate a more fluent reading of the text, hyphenation has been applied to the justified textblocks in order to avoid excessively large and irritating gaps. Those artificial hyphens are marked in grey, so they can be distinguished from those that were expressly written in the original manuscript. The folio specifications can be found on the left side next to the transcription of the original text.

THE TRANSLATION

A considerable amount of Wilhalm's text is based on Johannes Liechtenauer's teachings, which are recorded in verses (called 'the text') and commentaries (called 'the glosses'). Wilhalm's version is often a deviation or a deformed version of older sources, occasionally to the extent that the original meaning has been lost. At times, it is highly dubious whether the scribe actually understood what he was writing. Quite likely he did not entirely comprehend his source – and possibly had a hard time actually reading and deciphering portions of it. In comparison with older text versions, there are occasional outrageous misreadings. In the translation, I did not amend those atrocities; instead, I tried as best as I could to stick close to the original. This may result in some oddities and sentences that do not seem to make sense. Only extraordinarily blatant errors such as a mix-up between 'sein' and 'dein' ('his' and 'your') have been cautiously corrected, since they alter the meaning considerably and are certainly only due to inattention by the writer. Apart from that, Wilhalm – as well as his predecessor Liechtenauer – deliberately phrased their teachings in 'secret and obscure words'. I therefore frequently had to make a choice. Particularly obscure passages are provided with annotations in the footnotes. Master Liechtenauer's verses actually rhyme – at least in the early sources. Sometimes, they are so corrupted in Wilhalm's version that these rhymes disappear completely. In order to stay as close to the meaning as possible, they are rendered in prose in this edition. Wilhalm inserts the Notes into his text without distinguishing them in any way from the glosses. In order to make them clear to the modern reader, I have highlighted these verses in **bold** in the respective section (pp. 309–331).

Since both Wilhalm's and Liechtenauer's texts are quite ambiguous and since there is more than one way to translate a specific passage, I have occasionally sought refuge in different translation variations throughout the text to better explore the broad spectrum. Due to the complete absence of punctuation, I felt the need to resolve the multi-clause sentences in more easily digestible components. Frequently, I needed to make a choice, and I hope to have done justice to Jörg Wilhalm and what he intended to convey.

I have not translated the majority of the specific technical terms of the Early New High German original text. Some of these terms are deliberately ambiguous, possibly in order to give a special word more scope or to obscure its original meaning. Some more generic words (such as *absetzen* or *versetzen*), however, have received a translation (to set off, to displace) in order to facilitate a general understanding and not to burden the text with clumsy constructions such as 'perform an *Absetzen*'.

THE GLOSSARY

The German entries in the glossary follow the orthography of the original Early New High German text. Since the spelling in the source is not always uniform and a considerable number of variations occur, I have unified them and adapted the spelling slightly: capitalisation and separate spelling of words follow mostly modern standards. Behind the entry, the term can be found translated into modern English in brackets, as far as it can be reasonably transferred. Special technical expressions, however, appear only in their original version. Vocalic 'V' can be found under 'U' (see for instance 'Vbergen' which in modern German would appear as 'Übergehen'), the consonant version under 'V'. 'B' and 'P' are often used interchangeably, which is indicated by a corresponding reference. To make it easier to find a term, I have noted the respective folio numbers in the glossary behind the definition.

Jungk Ritter lere Gott lieb haben und frauen
Eren das frauen wol und bis manlich als
man dan sol ziert dich auch vor lastern
darzu las dich nit betriegen das ainis
manes mutt gen dem der die veracht
thutt und sich dem hin in ein herliche sach
und leer darzu manliche Riterschaft durch
~~schimpfen und schertzen~~ mit freuden geben
die stain stosser stangen schieben vechten
und ~~kinder~~ dantzen und springen stechen
und turnieren darmit sol man schone frauen
hofieren durch schimpfen und schertzen
vechten wil haben ~~lib~~ grosse welche
das schirket gen kain vechten sehr wil
treuen warum wer verlor die kunst
vom schwert gert der kunst und auch die
grose schwach machet die verachte fechter
wirt darum sitt man gar vil schlag
das die verachten oft werdent geschlagen
und ist ein verachter gen dem ein man
nitt vechten wil kumer hin und kunst
nit gelernet hatt das ~~klag~~ ich auch
Jörg wilhalm hutmacher von augs
purck früe und spatt

Wer auff dich vil hawen
erschrecken sol sich frawen
und wenn du hast versetzt
So seind all sein kunst geletzt

Die zwen stond all baid
Recht in dem oberhaw

Wiltu kunst schawen so sich linckh und recht mit hawen und linckhs mit rechtem ist das du starckh so Recht zuvechten wer nach eilt mit hawen bedarff sich kunst wenig frawen haw nachent wo du wilt kain durchwechsell kumpt an deinen schilt zu kopff zu leib das vechten nit vermeid mit ganczem leib dich was du starckh begerest zutreiben hernach was du schlechst frisch auff

Der stet im oberhaw

Das ist der zornhaw von der lincken achslen

Das ist der zornhaw auff der lincken seitten und der selben arm von der lincken achslen und haw im schwert hinein lanck zu dem kopff und verkher und wind damit und las ab der sterck lauffen und schlag aber wie vor geschriben stat und hinder sich hinweg

Der stat im oberhaw

Das ist der zornhaw von
der rechten achslen

Das ist vom zornhaw wer dir oberhawet so nim
den zornhaw starckh auff dein achslen und haw in
schlecht hinein uber die versatzung und volgend
und wind mit dem schwert als woltestu dich ver-
setzen und lass abfallen zu baiden seiten mit halben
vollen und schlach in mit der langen schneid nach
dem rechten or zu dem kopff hinein

Die zwen stond baid glinckh in der oberhaw

Vicht nit oben glinckh So du gerechts bist und ob
du glinckhs bist und im rechten auch ser hinckst
vor und nach die zway ding sind aller kunst brun-
sprung schwach und sterckh indes wart dem an
wait damit merckh so machstu lernen mit kunst
ern arwaitten und wer erschrickt gern der sol kain
vechten lernen

Das ist der zornortt　　　　Der stat im oberhaw

Das ist die aufrichtung und merckh vor die oberhawet das du im mit dem zornortt eindringest und weitt er das swert so nim es oben ab on gfar und bis starckh herwider wend haw und stich sicht er es so nim es mit das eben merckh haw stich schwech und sterckh und nims mit halben vallen und schlag aber hinein wie vor und hindersich hinwech

Der stat im oberhaw

Das ist der ander zornortt

Das ist der andr zornortt auff der andern seittn den man auch also wen sy gend baid von der zornhaw von der achsel und muss einer auf einem haw ain stich machen und muss den haw vergessen und das man auch flue zornortt wege das ob

Hie vahet an der krieg

Die zwen stand in dem krieg

Das ist die aufrichtung vom dem krieg vor des kriegs oben zwen der vacht oben under geschrenckt unde vor und nach dem krieg daz dir nit gach sein des in allen sachen wiltu den krieg machen dz hast merckh

Das stuckh ist in dem krieg

Das ist auch von dem krieg vnd solt auch wissen
das in allen winden haw vnd stich schreitter
vberlauffen schwert auswinden in allen kreffen
wiltu der maister üben wann es sind des vil
bruch darinne vnd auch darzu nim war wie
einer mit dem krieg geben stoss merckh

Das ist ein aufwinden und bruch von dem krieg

Das ist aber ain stuckh von dem krieg und ein bruch des vordern stuckh und ist ein aufwinden das gibt haw stich gut schniden finden auch soltu mit brauchen haw stich schnitt winden in allem treffen wiltu der maister üben

Das ist auch von dem krieg

Das ist auch von dem krieg schuech zu dem ort
und nim den hals one forcht auch soltu ver=
wenden krum und such umb und auff und
brieff im sein gfertt ob er sey waich oder hert
in sein gfertt glass merckh

Das ist auch von dem krieg

Das ist von dem krieg die aufrichtung wer also einen wartet bey dem halss mit dem schwert der mus wennen mit ein weich winden und wider winden das geit hut schwech und sterckh wider und verwend dem schwert an dem halss und zuckh den linken fuss hinder sich und schneid in ob dem halss bloss merckh

Das sind ötter in dem krieg

Das sind die valschen ötter in dem krieg das gar vnglichen zu weren warumb es gehörent vil durch
gengs dar zu vnd winden vnder durch oder wind
oben vber gehörent vnd wirfft den ort auff die
hend vnd werfen zwerchen durchlauff auch vil
künsten

Das ist aüch von dem krieg

Das ist der krieg wan du ainem schlechst in die rechten seiten
so nimpt der ander den krieg vnd bedarff dir nit mer
von dem bloss versetzen mit der langen schneid vnd so
schlecht er dir mit dem schwertt oben vber vnd thue
als dü es gmalt sichst

Das ist auch vor dem krieg ein schwertnemen

Bloss merckh das ist ein schwertnemen vnd get auß dem krieg das nimpt einer gar gering hinweg das es sind noch gar vil gutter stuckh die zu dem krieg gehörent vnd ist das kriegen manigerlay vnd darum so soltu merckhen gar vil ding

Hie vachten an die vier bloss

Das sind zwu bloss

Hie vacht an die bloss zum kopff und auch zu der
seiten die vier bloss weiss zu remen so schlagstu
gewiss on all gefar on allen zweivel wie er ge
bar das weistu wol gerat auch kumpt es noch
hinden hernach wie du solt die vier bloss
brechen bloss merckh

Das sind auch zwu bloss

Das sind die andern zwu bloss zu der andern seitn
darum so lug und merckh auff wie die gand
und wer die bloss anders nempt der wirt
oben bey dem ort geschennt bloss merckh
behenndigkh und streicht zc

Das ist der krumphaw auff
der lerchten seyten verwechsel

Der statt im ober
haw

Das ist von dem krumphaw die aufrichtung kum
auff behend wirff den ort auff die lenck hand
kum zu der schwöche wiltu den magst schwechsl
kum mit kurczhaw dein durchwechsel darnach
Oberhaw oder verwend wiltu winden und treib
im in der drianckel oloss werckh

Der stat im oberhaw

Das ist der krumphaw von der lincken seitten

Das ist krump auff behend wer wol versetzt mit schritten er teil haw letzt haw krum zu der flech en wiltu den maiste schwechen wan es klitzt oben so stand ab das wil ich loben krum schlecht und nit kurtzhaw den durchwechsel damit schaw haw krum wer dich jeret der edel krieg in verwirret

Das ist wie man die vier bloss
brechen sol der ein der schlecht
im mit der bloss in die seiten
So bricht der ander hie vor
und schlecht im hinden auff
Sein kopff

Der schlecht die bloss
dem andern in sein
rechten seiten

Das ist bloss brechen und wiltu dich lernen die vier
bloss kunstlichen brechen so du aber duplieren wider
recht mutieren ich sag dir vorwar sich schickt kain
man vor sam hastu vernomen zu schlagen mag er
kainer kumen das hab ich vernomen gleichs
merck

Da bricht diser die blöss
auff der andern seitten
und ist auch die blöss
brechen

Der schlecht die blöss
dem andern in sein
linckhen seitten

Das ist die blöss brechen auff der andern seitten
und brich sy wie du die gewalt sichst du must
auch oben duplieren und andere fecht mutie
ren Item die blöss brechen solt gewalt sein
darvorn vor dem krumphaw

Die hut vom tag — Der alber

Item das sind zwu hutten und ist das die hut
vom tag der sein schwert ubersich hebt und
der Ander der sein schwert undersich helt bei dem
lincken fuß das ist der alber und haist auch d
die eisin port dan es sind vier huten ob
und schrenck das sind auch zwu huten oben merck

Schaitelhaw

Ein leger oder die eisern port

Item das ist von der kron und der der das schwert obersich hebt der stet in dem schaitelhaw und der ander in ein leger und wan der schlecht von dach ab so nimpt der ander die kron alß schon stoß merck

Das ist der zwerchhaw

Der statt im obern haw

Das ist die zwerch be nimpt was von tach herkumpt die zwerch mit der sterkh dem swait damit wirkh die zwerch zu pflug zu dem ochsen gehört sich und er was sich wol zwercht mit springen dem haupt ob fert und was oder vernaw von andern auff zwinnich nach wünsch di kreutz glests wirkh

Das ist die zwerch auff der andern seitten

Der statt im ober haw

Das ist die zwerch auff der andern seitten den nim auch wie auff der ander seitt vnder zwerch durchlauff vnd auch ring vnd in den elenpogen in der wag vnd trit an den lincken vnd bis nit lass in baiden zwerchen magstu ringen blos merckh

Der bricht die zwerch

Der schlecht der die zwerch

Das ist das stuckh wie einer die zwerch brechen sol da bricht ein zwerch die andern mit auff sitzen vnd mit vnderschiessen vnd mitt oberschiessen daraus machstu ainmal die zwan wenden ob du wilt dan es sind zwai wenden die den man ser treffen das merckh

Der stet in der hůt vom
tag und weret sich als
vil er mag

Der spricht also was sich
wol zwerecht mit sprüngen
dem haupt geferet das
ist das stůckh

Das ist auch ain stůckh von der zwerch
was sich wol zwercht mit springen dem
haupt gefert auff ain seitten zů der andern
ist auch eins das ist besser den das also
wunnen gloss merckh

Der leit in der zwerch vnd hat di
genomen mit ein sprung vnd
stost im den arm hinnach von
im

Das ist auch von der zwerch das nimpt man
auch mit ein sprung das nim also wie di
es gemalt sichst darnach nimpt der bruch der
im sich bruch wie dan es ist kain stuckh es
hat ein bruch darum merkh kunst wer
nu be stan

Der statt auch in der zwerch
und der will zwerch stossen

Der aber bricht im
und zwercht sich ob
uber sein kopff

Das ist der bruch darwider du magst
wol fluchs ein schrit vorwerts und hengest
das er dich nit erlauffen mag den ein stuck
bricht das ander und merck dabei was dir
gut sey

Das ist auch von der zwerch mit
springen dem haubt geferett
wie er gefan mit seiner kar
er als gefan

Das ist auch ein stuckh was sich vil zwerch
mit springen dem haupt geferet und ein schnit
kain sich und uberleiten und bey dem halß warten
also verstand es wie du es aber gemalt sichst
gloss merckh

Das ist die zwerch oben ein auch ein stuckh

Das ist aber von der zwerch von dem vor
gen das du es verstandest das gar vil auß
der zwerch kumpt wen man es sucht wen
Es ist gut zu allen solchen schwerch und
streich machen Also merckh

Das ist der ochß — Das ist der pflůg

Das ist von dem ochsen und pflůg die aufsůcht
auch die hůt und die zwai hůt sind aller kunst
ein ursprung ochß und pflůg sind der kunst hart
stuck In allen sachen den alten schaidt nit zu
machen so bistu sicher in allen dingen und
mag dir nit misselingen

Das ist der ochs Das ist der pflug

Das ist der ander ochs und pflug und sind auch den kriegs hutten sind Item der ochsen und pflug sol man treiben von baiden seiten dan sy haben vier winden und vier ort die magstu brauchen wie du wilt wans sind auch zwen hutten [illegible] merckh

Der statt gewapent

Das ist auch von dem schilhen und hat in den halss so nympt der ander ain gewapent stückh und bricht Im sein stückh als du es gmalt sichst das ist auch ein stückh ein bruch bloss merckh

Das ist auch von dem schilhe stuckh und
bruch und der da hat das schwert auff
dem halß das ist der bruch und das
ander das stuckh das merckh wie du
es gemalt sichst

Das ist aber vom schilchen der bruch das
uberlauffen das bricht ein der schilch
starckh schlag oder schnit und zwerch
fluchs und bald und darumb ist kain sterch
es ist ain bruch darauff gleß merckh

Das ist ain ander stuckh von dem schilher
Schilch zu dem ort und ainn den halss
one forcht und dolbri so merckh und
lern kunst das du dich darauss kundest
vechten ploss merckh

Der statt schlecht in der ober versatzung

Das ist der schilcher der ein bricht was buffel schlecht oder sticht

Der schilcher einbricht was buffel schlecht oder sticht wer wechsel treibt mit schilcher darauß beraubt schlecht er dich kurtz an dem durchwechsel Schilt im an schilch zu dem ort und nim den hals one forcht schlecht zu dem obern haubt behend wiltu best-
weren und rüder sterr

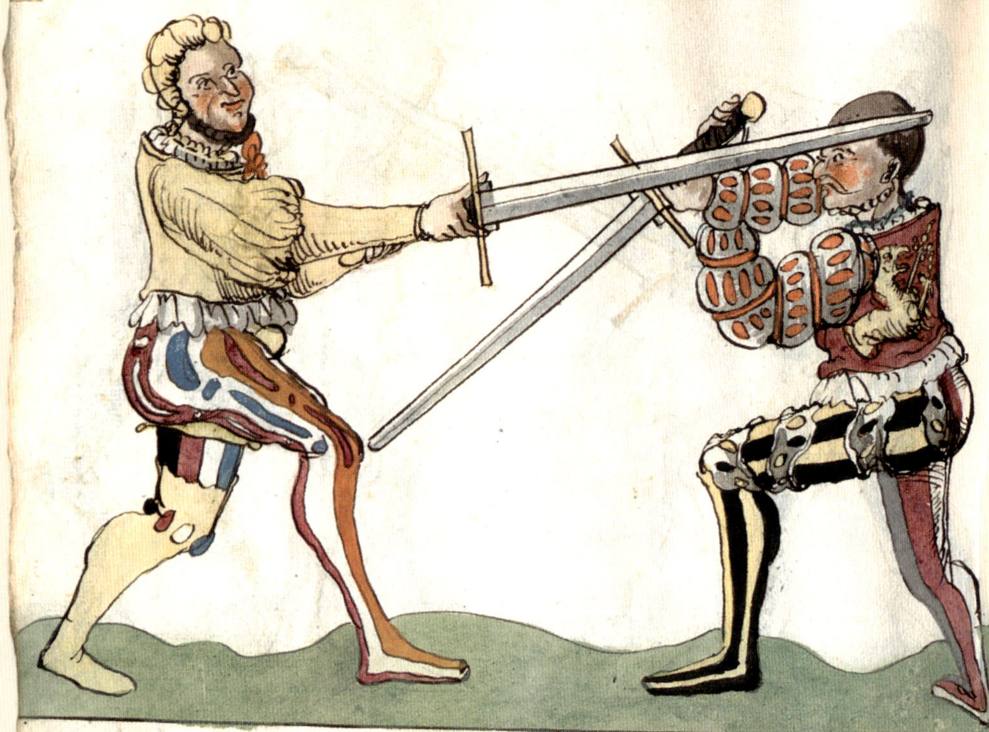

Der stat schaitler in der ober versatzung

Das ist der schaitler

Der schaitler dem antlitz ist gevar mit
seiner ker der brust fast vierwar was
von im kompt die kron das abnimpt der
schnidt durch die kron so brichst sy hart
und stos die straich druckh mit schnidten
sy abdruckh in allen sachen lass dich frisch
kurtz machen

Das ist schleich zu dem ort
vnd nim den halß one forcht
vnd darzu sterckh in der waff

So spricht der bös gesell
gar vil brich der wadel
die noch hernach komen

Das ist vom stechlichen schleich zu dem ort vnd
nim den hals one forcht vnd stel dich in die
waf vnd halt fast an dich vnd wan du von
im wilt so wend im aus vnd hinderst
flaß merckh

Der schlecht oben uber

Der hatt den hals one forcht

Das ist auch von dem schilch und ist ein
bruch daraüff wie du es gemalt sichst und
nim war das noch mer stuckh sind die
es brechen und nim eben war wie er
gebar floss merckh

Das ist der schaittelhaw

Das ist der stich

Das ist von dem schaittelhaw der aufrichtung
das ist schlecht mit schaittler schittlet dem antlich ist
gevar mit seiner kar der brust vast gevar
und was von im kompt die kron also schon
glesch merck

Das ist die kruch schneid

Schaitelhaw

Das ist auch vom schaitelhaw es ist ze wissen
das einer den schaitelhaw sol nemen mit dreyn
dritten und einer mag in nemen mit ain drit
der vor im kumpt die kron das abnympt der
schnidt durch die kron so truckstu in hart und
schon die straich druckh mit schritten in abzuckh
floss merckh

Das ein versetzen

Das ist ein leger

Das ist von den vier legern und auch alwegen
ain leger und ein versetzen darauff die vier
leger allain davon holt die sach und fleuch
die gemain ochs und pflug aber vom tag ist
wol sein hut

Das ist auch vor den letzern und ist zwo=
rissen das ein letzer das ander bricht und
das sind zwai letzer ainer leitt oben der
ander unden das merckh also und wer
nim das der underst dem obern wol duch
was gen und gibent im auff sein kopff
und darum durchgen thut woll so ich
hatten sol floss merckh

Das ist auch von den letzern das der dem
aüff dem letzer schwert kumpt auff vnd
schlecht im vber sein ~~hand~~ schwertt vnd
list das vber schiessen auch ein letzer das
ander bricht bloss merckh

Das ist ein versetzen

Das ist auch ein leger

Das ist von den vier versetzen die die
leger ser letzen doch vor versetzen hüt
dich geschicht es dir nott es mutt dich
und ob du versetzt und kum vor hastu ver
nomen zu kainer kunst mag er nitt
kümen bloss merken

Das ist aber von den vier legern und
die versetzen darauff und uberschiessen
das sind auch leger das merckh da von
nym mit der sterckh und lauff in der
wag und wen er dich wellt aufheben
so mach im ein schnidt wenen ob du wilt
oder mocht verkeren was dir am besten
wirtt ploss merckh

Das ist aber von den vier lesern vnd von den versetzen da merckh auff vnd schaw des treib auch zwai leder stand das ist auch ain leder der vberschossig hurt vnd haben die leder ein end stand die vier versetzen Sy sind als bey ainander dan es ist ir vil der sy sucht also merckh

Das ist von dem nachraisen leer zwiuach
in der were vnd zway daraus ainer vnd
den antwurt daraus besind vnd briff
dem schwert ob dÿ sind waich oder hört
das selb leer Judas des wort schwindet
Drei nachraisen zwiuach so nÿmstu den
alten schnidt mit macht vnd hab acht
gleich merckh

Das ist ein anders nachraisen den
das nachraisen das soltu nemen auff
baiden seitten drit lanck wiltu schreitten
wen das nachraisen ist manigerlay
und das ist das nachraisen wan einer dich
uberschlachen hott und will von dir gen oder
abziechen So volg im nach und wer
dich bloss merckh

Das ist von dem überlauffen da merckh das das überlauffen ist so einer den andern aussen überlaufft mit dem knopff auswendig mit machst und das ist ein ausser überlauffen und der ist noch einer und darum wan einer dir will überlauffen so heb dein schwertt grachlich in die höch so kan er dir nit über lauffen glosa merckh

Das ist auch von dem überlauffen und das
ist ein zwerendichs überlauffen und wer nur
hewtt und wer die nider überlaufft so wirt er
geschemmtt und wer es glaubt oben so setz es
ab das will ich loben dein arwait mach waich
oder hertt und durch das zwaitlich wie du
wilt gloss merckh

Das ist auch ein überlauffen und ist gewapp
nett und get ein kurtzen damit und darum
So merckh eben auff wan er will ein werffen
oder den klincken schiss und gehört zu dem
überlauffen vil eben man suchen will
gleich merckh als es oben gemalet statt

Das ist auch ein überlauffen mit dem
knopff und kin überlauffen gewappet
mit schencken wür vor und durch zu fend
wirff in damit und thu als es oben
gemelt statt bloss merkh also die
überlauffen

Das ist von dem absetzen wer auff dich
haut von oberhalb So setz im ab mitt
dem schwert mit der kurtzen schneid sein
haw ist er waich so lass umb dein kopff
gen und schlag in mit der zwerch ist
er aber hertt so schlag nit und nim
und verwend im an dem schwertt und
stoss merckh

Ein abschern

Das ist ein abschern auff der andern
Seytten und das treybt man auch
mit der kurtzen schneyd wie vor und
merckh auff die schwech und sterckh ob
er sey waych oder hertt und lass merckh

Ain absetzen

Der ligtt im langen ortt

Das ist von dem absetzen zwais stich schnelig
kurtzen letzen und der hinder ligt in dem
langen ortt und der vorder setzt im ab
mit ain absetzen

Dÿn absetzen Dÿn lanng ortt

Das ist ain ander absetzen vnnd weder setzt dem ander den ortt ab glosa merck

Ein absetzen   Eyn langort

Das ist wer will absetzen hew stich schnelliglichen
letzen vnnd wer auff ~~sticht~~ wil so lug das dein
ort trifft vnd dem sein bricht von baiden seitten
trifft woltu schreitten das ist auch ein ab
setzen stoss merckh

Ein absetzen

Ein langer ort

Das ist ain vil absetzen daran stich kum
schwerlich blaichen lassen und ist auch dis
absetzen gloss unwirckh als Es oben gmelt
Statt

Das ist der erst oberschnidt uber die arm
und schnaid ab die hörttung schnel und flüc
mitt dem gschritt und hab acht und nim
den schnidt mit macht und also thu closs
merckh als es ~~[illegible]~~ oben gmalt statt

Das ist der ander oberschnidt den man auch
mit der lincken stemen schloss merckh

Das ist von dem andern schnitt der drit
der schnid ab die hörttung durch baid arm
das merckh als es oben gmalt statt vnd
nim das schnell vnd bald hinwegk stoss
merckh

Das ist der viertt schnidt den nim auch wie
du jn vnden gmalt vinst vnd nim in
starckh oder warch wie du willt vnd flech
hinwerts bald vnd behend das dir dein schnidt
volendt flosse merckh

Das ist auch ain abschnidt den nim als du
in oben gwalt sichst und schnid im ab die
hertung schnell hand flux und tring das er
dich nit erfolg mit dem nachraisen also
merckh

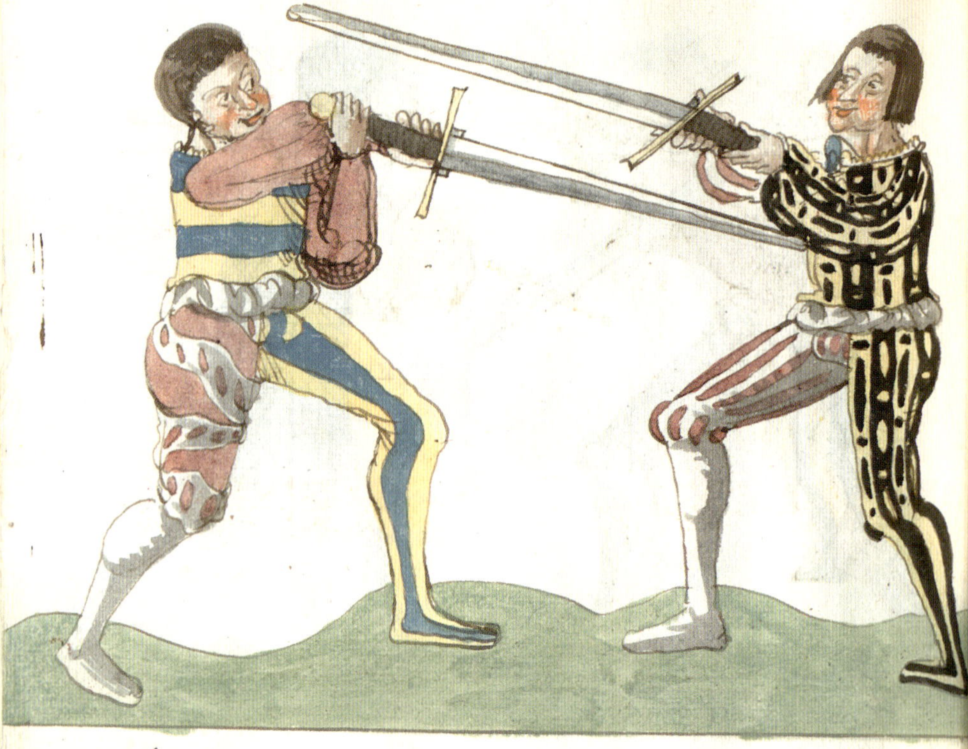

Item das ist der Alt schnidt den treibt der vorder
den man also und herwertz zu unden auff durch
den fechtan einem starcken und langk fechter
und fleucht keiner bloss wercken

Das ist der ~~altt~~ ander altt schnidt zu der
andern Seyttel den man also wie er oben
gemalt statt vnd man in flechauff vnd
hindersich hinwegt das er dich nitt erlangt
ploss merckh

Das ist von dem hend truckhen das schwert durckh in die hend und thů das schwerd dem schwert wend zů fläch durch die hend stoss machen

Das ist auch ein hend trucken das man also
vnd Inwerffen mein dein schwertt vnd schlag
Im nach der blösse vnd truckh Im von dir
starckh stoss Inwerffen darauff vnd man
kan wie er gethan

Das ist von dem hengen zu zway hengen vnnd
auß ainer hand von der andern In allem fecht
hauw stich leger waich oder hertt mit sein treff
das ist das hengen das sich zwerchst als du
es oben vnnd vnden süchst stoss merckh das ist
ain hengen auff der ainen Seitten das ver-
stannd eben

Item das ist ain anders wenden auff der andern
seitten das thůt er mit dem kurtzen schwertt
vnd heb den spitz bas vndersich wie dü es
oben gemalt sichst vnd sind die zway stuck
zwu gůt versatzung zu allen sachen gloss
merck

Der bruchfenster

Item das ist das sprechfenster bruch vnd merckh
Stand frolich vnd sich sein sach dein zwerch
Mit der sterckh vnd kreind in wenden
durch vnd merckh als du es oben gemalt
sichst &c stoss merckh

Ain langer ortt

Das ist von den vier valschen örttern und
ist der ainen ain langer ort und sy genn
zu baiden seitten da merck auff das in der
lang ortt in das gsicht gatt und ist gut
zu allen sachen ain krieg damit zemachen glosse
merck als du es oben gemalt vindest

Das ist der ander verborgen ort den nim
als du in dem gmalt sichst der ort auch
zu dem gsicht darumb so merckh auff vnd
such was gut sei darauff

Das ist der stich verborgen ortt und ist ein
valscher ortt den merckh als er oben gmaltt
statt du magst auch wenen das durchgen
oder durchwechsel wie du wilt das durch
gen wen also gee nu mit der spitzen und
durch bloss merckh

Das ist der viert valsch ort den nim auch
also du in oben gewalt sichst vnd er dringt
dem das schwert von der hend vnd ist auch
ain guter ort vnd nim war vnd
merckh darauff Als du es oben gewalt
sichst

Das ist aber ein valsche ortt und ein valsche
hut den wan als du in oben gwalt sichst
und wan er auff dein schwert valt
so wind auff mit dein schwert so stost du
in der langen verzaschungs stoss merckh

Der stat im einhorn

Das ist auch ein valscher ort und gert
auss dem sprechfenster und den nim
als du es oben gemalt siechst und ist
das einhorn clos unverletz

Item das ist die kron der sein schwert gewappent helt
in der hand schon und nim war schon mit der
kron die ist gutt zu alle ort zu allen worten und
ist starkh darin oder schwach wie du wilt schlos
merkh

Das ist der schnidt durch die kron und den
arm also stan als du es gemalt sichst in
allen sachen wiltu ein schnitt machen
also mercks stan auff den schnidt in der
kron

Item das ist der bruch auff die kron da merckh also
schon als du es gemalt sichst vnd der das schwert
vbersich hebt der bricht die kron mitt dem
kreussen am schwertt mit dem bruch da lug auff
seiner seit nim die kron hinweg also merckh

Hie hatt der lang zettell ein end das helff
vns gott also behend vnd ist die kron des
Wallers letst vnd ist das aller best

Ein gut gemain ler des langen schwerts

Wiltu kunst schawen so sich linckh und rechts
hawen und linckh mit rechten ist das du starckh
gefrest zu vechten wer nach geit hawen der
darff sich kunst wenig frawen haw im nahent
nach was du wilt kain durchwechsel kumpt an
deinen schilt zu kopff zu leib sicht nit vermeid
mit gantzem leib vicht was du starckh gefrest
zetreiben hernach was du schlechst auch nit ober
linckh so du rechts bist und ob du linckh bist im
rechten auch ser hinckt vor und nach die zwai
ding sind aller kunst ain ursprunckh schwach und
starckh indes merck dein arbait damit merckh
so magstu lernen mit kunst arbaitten und
wer erschrickt gern kain vechten sol er lernen
kunst hat lenge von der rechten hand gerne wirt
du gewar wie glauben zu kunsten gern zu
kunnen

Das ist der text und die auslegung der schaitler

Zornhaw krumphaw zwerch durch haw schilcher mit
schaitler vorlauffen haw sticht durch wechsel auff
aller abschneiden hand drucken hengen winden
mit blossem schwert schlachen staich stich mitt
stossen

Das ist die ausrichtung vom zornhaw

Wiltu mercken wer dir oberhawet zornort im
dratt heraus er wind stich sicht er es so nim es wider
das soll merck haw stich leg waich oder hert
indes vor und nach dem krieg sei dir nit gach
wer des kriegs unden bemerckt der wirt oben
beschemet In allen winden haw und stich

Schneide finder auch so solt du nit meiden hau sich
schneide zuckh in alen treffen wiltu den maister
~~effen~~ etc Das ist von den vier blössen

Die vier blöss weiss zu remen so schlechst gewiss
on alle far on alle zweivel wie er gebar hastu es
vernomen zu kain schlag mag er komen ~~etc~~

Das ist wie man die vier bloss brechen sol

Wiltu dich rechen die vier bloss kunstlich brechen
du oben duplier unden zu recht mutier ich sag
dir fur war sich schirmbt kain man one far hastu
vernomen zu schlagen mag er kaum komen das
hon ich vernomen

Das ist die aufrichtung vom krumphaw

Krumphaw auff behend wirff den ort auff die
hend wer wol versetzt mit schreitten er vil haw lest
haw krump zu den flechen wiltu den maister schwechen
wan es klitzt oben so stee ab das wil ich loben krump
und schlecht und kurtzhaw dem durch wechsel damit
schaw haw krump wer dich irrt der edel krieg
in verwirrt das er nit waist wie wer wie er
gebar etc

Die aufrichtung vom zwerchhaw

Die zwerchhaw benympt was vom tag herkompt
die zwerch mit der sterckhe dein arbait damit
mercke ochs pflueg alber vom tag ist wol sein feind
was sich wol zwercht mit springen dem haubt
gefert was wer verfiertt unden nach
wunsch fehlerant verkerer zwinckt durchlauff
damit dringt nim den elenpagen greiff

Springt im wol in die wage mit der glingken ellen //
// bogen zwerchart mit macht und das schreitt in linkh
und bis nit lass

Die ausrichtung vom schilhau

Schilch ein bricht was buffel schlecht oder sticht wer
wechsel trewt mit schilch nimbt er darauss beraubt
schilch er dich kurtzhau dem durch wechsel sigt im
an schilch zu dem ort und nim den halss on vorcht
schilch zu dem haubt oben wiltu in beleiben

Die ausrichtung von dem schaitelhau

Der schaitler dem antlitz ist gefar mit seiner ker
der brust vast vierwar was von im kumpt der die
kron das abnimpt schnaid durch die kron so brichstu sy
hart und schau die straich drukh durch mit schnaiden
sy abziehe

Die ausrichtung von den vier legern

Vier leger allain davon halt man und
fleuch die gemain ochs pflueg alber vom tag ist
wol sein fueg

Die ausrichtung von den vier versetzen

Vier sind der versetzen und die leger auch sere letzen
doch vor versetzen hiet dich geschicht es dir not es
müet dich ob dir versetzt ist und wie das derkumen
ist hör was ich dir rat kum hin und nit zu spot
streich auff hau schnel mit dritten selban vier
ender und blib darauff so lernstu binden

Von dem nachraisen

Nachraisen lere zwiuach schnaidet in der were
zway darauss nim dein arwait darauss beschirm
und brüeff sein gfert sein leger waich oder hert der
selbig lere jedes das wort schnidest sere nach //
// raisen zwiuach nim du den alten schnit mit macht

Die aufrichtung von dem oberlauffen

Der under remet und oberlaufft der wirt aber geschemett wen es klitzt oben so sterck das will ich loben dein arwait mach waich oder hertt und drucke das zwingen

Die aufrichtung von dem absetzen

Lere absetzen hau stich kunstlich wer auff dich sticht so lueg das dein ort trifft und sein bricht von baiden seitten triff alle mal wiltu schreitten

Von dem durchwechsel

Der durchwechsel lere von baiden seitten stich mitt sere und wer auff dich sticht bündett durchwechsel in schier windett

Die aufrichtung von dem treffen und zucken

Trit wen die binden zucken gibt gute funde zuckt er so zuck mer zuck auf allen treffen wiltu den maister effen

Die aufrichtung von dem durchlauffen

Durchlauff las hangen mit dem knopff und greiff wiltu ringen und wer gegen dir sterckt sterckher durch lauff damit merckhe

Die aufrichtung von dem abschneiden

Schneidt ab die hertte von unden in baiden geferte vier sind der schnidt zwen unden zwen oben mitt

Von dem hend truckhen

Das schwert kerd zu der flechen truckh in die hend von dir dein schneid wend zu der flech durch die hend

Von den zwaien hengen die aufrichtung

Zwai hengen werden auss einer hand von der erden in allem gefert hau stich leger waich oder hört in iren gefert

Von dem sprechfenster die ausrichtung

Das sprechfenster machen / steh frolich und sich sein sache war sich vor dir zucht ab schlachen das er schnabe schnitt er sich gegen dir streichen schlagen das er rise

Das ist die ausrichtung und die beschliessung der gantzen kunst war wol sicht und endlich bricht es bricht etlichs in dreyen winden war recht wol hengt und windet da mit es bringt warhet mit rechtem betrachten und einer der da windet und 24 sind der winden und die zal mitt verfassung von baiden seitten acht winden wer mit schreitten treit und brufft der schritte ab sie und wach oder dritte also endet sich maister liechtnauers kunst des langen schwerts

Oberhaw zornhaw zornortt der kruphaw die vier blöss krumphaw zwerchhaw eyle und pfleug schilcherhaw schaittlerhaw die vier leger die vier versetzen das nach raisen uberlauffen absetzen abschnieden zwai hengen das sprech fenste die kron die sturtz des schwerts Johannes liechtnawers

Die siben maister haw

Der zornhaw der krumphaw der schilhaw der kreishaw der zwerchschilhaw der schaittelhaw der kronhaw

Das ist der text von den vier winden das ist ein haw ein stich ein schnit ein streich und ist auch der text von dem langen zettel des langen schwertts des lichtenauers kunst und ausrichtung also

Item hie merck wie du auf den zwaien obern hengen sollt treiben das ist auf den ochsen soltu treiben acht winden von baiden seitten von der rechten seitten 2 von der lincken seiten 2 und treib die also wen du mit dem rechten zu im kumpst so stand vor mit der rechten seitten in dem ochsen hawt er dir den oben ein zu deiner lincken seitten so wind gegen seinem haw die kurtz schneid an sein schwert auf den ochsen und stich im oben ein zu dem gesicht das ist ein winden versetzt er den stich so bleib am schwert und wind wider auff auf der rechten seitten aber in den ochsen die kurtz schneid an sein schwert und stich im oben ein zu dem gesicht das sind zwai winden am schwert auf dem oben hengen vor der rechten seitten Item das ander ober hengen treib also wen du mit dem zufechten kumpst so stand vor deiner lincken seitten in dem ochsen hawt er dir den oben zu deiner rechten seitten so wind gegen seinem haw die lang schneid an sein schwert und stich im oben ein zu dem gesicht das ist ein winden versetzt er den stich so bleib an dem schwert und wind wider auff die linck seitten in den ochsen die lang schneid an sein schwert und im ochsen oben ein zu dem gesicht das sind die vier winden auf den obern zwaien hengen Item nun soltu wissen das auf dem pflug von baiden seitten das sind die zwai ~~hengen~~ under hengen auch soltu treiben vier winden mit allen iren geferte als auf den obern hengen also werden der winden achte und merck als oft du windest so ge=

// Merckh in ein zettlich winden besunder an den haw
und an den stich und an die schnidt also kumestu auch
den achtenden winden zu / und wie du haw und
stich schnidt treiben solt das vindestu alles darnach
in den stuckhen geschriben auch soltu vor allen
dingen die 8 winden wol gelerhen treiben lernen
mit schreiten zu baiden seitten und merckh
als bald du windest so soltu in einem yettlichen
winden besunder brüeffen mit wer den die zwai
gefert ob sy sind waich oder hört am schwert ist
darnach treib die stuckh als sy dir davor in den
nechsten vorgeklert sind worden

Das ist der text von dem schilcher

Lere wen du mit dem zu vechten zu gangest so
soltu schilhen mit dem gesicht ober kurcz gegen dir
vecht das soltu bei dem erkennen wanner dich zu hawt
recht langt er den die arm mit dem schwert hin
mit lang vor im so ist das schwert vor im kurcz
leistu den vor im in der hut aber weil er dir mit
dem schwert krauch vallen so ist es aber verkürczt
legt er sich den gegen dir in die hut des ochsen oder
des pflugs so ist es aber verkürczt auch was des alls
winden mit dem schwert vor dem man kurcz sind
und verkürczen das schwert und alle die also gegen
dir vechten den verchsel frölich durch auf hawen und
durch auf stechen mit dem langen ort damit zwinge-
// stu sy das sy das müessen versehen und sich schlachen
lassen schlecht zu dem ort und nim den hals ver-
vorcht elosz merckh das ist ein bruch wider
den langen ort mit einer betriebnus des gesichts

den leib also wen du mit dem zwerchen kumst statt
er dan und helt den ort gegen dem gsicht oder brust
so halt das schwert in dem zwerchen ochsen und schilch
mit dem gsicht zu dem ort und thu als du in darzu
hauen wollest und hau starck auß dem schilch
mit der kurtzen schnid an sein schwertt und schaib
in den ortt damit lang im zu dem hals mit ein
tritt des rechten fuß das ist der text schilch zu dem
oben haupt behend wiltu bedreben also merck
das ist ain ander bruch wan er gegen dir stet in dem
langen ortt wiltu in auff die hand schlahen so
schilch ain mit dem gsicht zu dem haupt und thu
als du in darauff schlahen wollest und schlag
in auß dem schilch mit dem ortt auff die hand

Das ist der text von dem schaittelhaw

Der schaitler dem antlitz ist gefar mit siner ker
der brust vast ist er var was von im kumpt die
kron das abnimpt der schnidt durch die kron so
brichstu sie hart und schnid die starck durch
mit schnidten so abzuck glose den schaitler bricht
die hutt alber und ist dem antlitz gefar und der
brust mit siner ker das merck den leib also wen
du mit dem zwerchen kumst legt er sich gegen dir
in die alber so setz den linken fuß vor und halt
dein schwert mit aufgerackten armen hoch ober
dein haupt in der hutt vom tag und spring zu
im und haw mit dem langen schwert schnid starck
von oben nider und bleib mit den armen hoch und
das im gehultz an seinem schwert band ober sich ston
das haist die kron und vertt damit auff und stoß
im den ort so wend dein schwert und stoß die kron
durch mit dem schnidt under dem knopf zu seinem
arm und truck also ist die kron wider gebrochen

vnd mit dem vnder kurtzen schneid vnd zeuch dich
damit ab

Das ist der text von den legern oder hütten

Vier leger allain darvon halt vnd fleuch die gemain
ochs vnd pflug alber vom tag sÿ dir nit vnmer Glosa
Merck die vier leger das sind die vier hutten do du
auß vechten solt die erst hutt haist der ochs da schick
dich also mit stee mit dem lincken fuß vor vnd
halt dein schwert neben deiner rechten seitten mit
dem gehültz vor dem haupt das die kurtz schneid gegen
dir stand vnd im den ort gegen dem gesicht Item
zu deiner lincken seitten schick dich also in den ochsen
stand mit dem rechten fuß vor vnd halt dein schwert
neben deiner lincken seitten mit dem gehültz vor dem
haubt das die lang schneid gegen dir stee vnd im
mit dem ort gegen dem gesicht das ist der ochs
zu baiden seitten

Item die ander hut die haist der pflug so schick dich also
mit sich den lincken fuß vor vnd halt dein schwert
mit verkerten henden mit dem knopff vndersich
neben der rechten seitten zu der huff vnd das die
kurtz schneid oben sey vnd den ort gegen dem gesicht
Item zu der lincken seitten schick dich also in den pflug sich
den rechten fuß vor vnd halt das schwert neben
der lincken seitten mit dem knopff vndersich der
huff vnd das die lang schneid oben sey vnd der ort
im gegen dem gesicht das ist der pflug zu baiden
seitten

Item die dritt haist der alber da schick dich also mit
sich den rechten fuß vor vnd halt dein schwert
mit ~~[illegible]~~ dem rechten arm
vor dir mit dem ort schloß auff der erden vnd das

die kurcz schneid ober gewandt sey

Item die vierd hut haist vom tag da schick dich
also mit und setz dein lincken fůß vor und halt
dein schwert mitt aufgeracktem armen hoch
uber dein haubt mit dem spitz und ste also in
der hut

Das ist der text von vier versetzen

Vier send der versetzen die die maister oder leger auch
sere letzen vor versetzen hüt dich geschicht es dir
es müet dich gross merckh du hast vor gelert
von den hutten so soltu nun wissen das man
nit versetzen die vier hutten brechen weis das
die kron versetzen nit darzu gehört wan es
sind vier hew die sy brechen

Item der erst haw ist der krumphaw der bricht
die hut auf dem ochsen Item der ander haw ist die
zwerch bricht die hut vom tag Item der dritt
haw ist der schilcher der bricht die hut die da haist
der pflug Item der vierd haw ist der schaitler der
bricht die hut die da haist der alber und wie du
vier hut mit den hewen brechen solt das wirdt
dir hernach zu den hewen geschriben darum so
hiet dich das du nit vil versetzest wiltu
anderst nit geschlagen werden

Das ist der text von dem versetzen

Ob dir versetzt ist und wie das dar kumen ist hör was
ich dir rate reiss ab haw schnel mit dritten gloss
merckh ist das dir ainer versetzt hat und wil sich
vom schwert nit abziehen und maint er wol
dich zu kainem stuckh lassen kumen so thu als
ob du dich vom schwert wellest ziehen und

zucke dein schwertt an dich bis an das halbtail der
klingen und var damitt bald am schwertt auff und
hau im schnell mit dem dupplieren oder mit der
kurczen schneid zu dem kopff gegen seinem ortt als ob
du im wolest nemen und bleib am schwertt und
hau im schlecht an die klingen mit der langen schneid
widerumb zu dem kopff

Das ist der text von den vier ansetzen

Secz an vier enden bleib darauff lere wiltu old mocht
enden das ansetzen ist ein recht stuck wann gatt
nach den vier blössen und gehortt zu treiben wen
du bald und weit geben Item das ansetzen treib.
also wann er mit dem schwertt mit dem rechten zu
im kumpt so secz dich mit dem schwertt in die hutt
des ochsens oder des pflugs weil er dir dan oben ein
hauen oder stechen vor seiner rechten seitten so kom
vor mit der versatzung und streich im den langen
ort ein zu der blöss seiner linkken seitten und wartt
ob du im anamittest setzen oder hawt er dir oben ein
von seiner linkken seitten so streich im den ort ein zu
der blöss seiner rechten seitten Item wann er dir
zu hawt von unden auff von seiner rechten seitten
So streich im den ort ein und setz im an das selbig
glaichen thu auch wann er von der linkken seitten
unden auff hawt wartt er den das ansetzen so ver-
setz und bleib mit dem schwertt auff dem seinen
und arbait behendiklich zu dem nechsten blöss
So lasstu in zu kainer kunst kumen

Das ist der text von dem Nachraisen

Nachraisen lere zwifach oder schneid in die wer zwai
aussere arm dein arbait darnach beginn und
prueff die gfertt ob sy sind waich oder hertt

Hie merckh der nachraisen ist vil und mangerlay und
gebrauch die zu treiben mit grosser fürsichtikait
gegen den vechtern die da auswarten und in langen
schwären vechten und sunst von Rechter kunst des
Schwertts nit halten

Item das erst stuckh aus dem nachraisen treib also wan
du mit dem zu vechten kumbst so setz den lingken
fus vor und ster in der hutt vom tag und sich gar eben
was er treiben will hawet er dir oben lingkhs ein
von seiner achsel das versich im nit und wart das er
dich mit dem haw nit uberlangt und ruckh nach
die weil sein schwertt mit dem haw undersich gett
gegen der erden so spring im zu der bloss seiner
rechten seitten ee das er mit seinem schwertt
auf kumpt so ist er geschlagen

Das stuckh haist die behemisch port

Item wan er sich verhawet und du im nach raisest fert
er dan bald auff mit dem schwertt und versetzt so bleib
mit der langen schneid starckh auff dem schwertt helt
er dan mit dem schwertt ubersich so spring mit dem
lingken fus hinder sein vechten und schlag in
mit der zwerch oder sunst zu dem kopff seiner rechten
seitten und arbait widerumb zu seiner lingken
mit dem duplieren oder sunst mit andern stuckhen
darnach als du empfindest ober hertt oder waich im
Schwertt ist

Das ist ein ander stuckh

Item wan er sich vor dir verhawet und du im nach
rawest bindestu daruff sein schwertt gegen seiner
lingken seitten schlecht er dan auf der versatzung bald
und mit der zwerch dir zu der rechten seitten so kum
im under mit der zwerch vorn under sein schwertt
gegen seiner lingken seitten an sein hals oder spring
mit dem lingken fus auf sein rechte seitten und haw
seinen haw auch nach zu seiner rechten seitten oder

treib den schnidt vber seinen arm zu dem kopff

Hie merckh ein gutt nachraysen

Item wen du gegen im vichst auf dem vnderhawen oder auf der strangkhut oder sunst gegen im in der hutt velt aber er mit dem schwertt auff das dem etwa du aufkomest des bleib vnden am schwert vnd heb vber sich weil er dir den auf versaczung ein hawen oder am schwertt einwindt so las im vom schwertt nit abkomen vnd volg im daran nach vnd arwartt damit zu seiner blöss am nachsten Item merckh du solt auf allen hutten vnd mit allen hutten nachraysen als bald du ~~sich~~ kommest wen er sich vor dir verhawet oder emplost mit dem schwertt

Das ist der text von dem empfinden vnd von dem wort das da haist Indes

Das empfinden lere Indes das wort schnidet sere Glosa Merckh das im schwert das fulen vnd das wort Indes die grösten kunst ist vnd wer ein maister ist vnd sein will vnd kan er das empfinden nit fullen vnd vernimpt darzu nit das wort Indes so ist er nit ein maister sunder er ist ein buffel des schwerts darumb so soltu vor allen sachen die zway ding wol lern merckhen

Item das empfinden merckh also wen du mit ainem zu vechten kumst vnd einer dem andern an das schwertt bindt so soltu indem als die schwertt zu samen glitzen zu hand vullen ob er waich oder hortt sey vnd die angebunden hat vnd als bald du das empfindest so gedenckh an das wort Indes das ist das du in dem selbigen empfindest behendigkleich soltu erwarten mit dem schwertt so wirt er geschlagen ee er sein gewar wirtt

Item hie soltu anmerckhen das das empfinden vnd das wort Indes eins on das ander nit mag sten [illegible] warumb also wen du im an sein schwertt bindest so mustu mit dem wort Indes vullen waich oder hortt vnd wen du gehast gfult so mustu aber mit Indes arbaiten also sind sy alweg beyeinander wen das wort Indes ist in

Aller stucken das ramen also Indes laufft durchzu
den duplieren und Indes mutiert Indes verke-
sett durch Indes laufft durch Indes nympt das
swert Indes schutt was du treib begert Indes
ist ein scharpff wort damit all maister des schwertz
verschnitten werden ~~wort~~

Das ist der text

Nachraisen zwifach trifft man den alten schnit mit
macht Glosa merck ist das du die nachraisen solt haben
zu baiden seiten und schlach auch darein kunsten und
das ramen also wen er sich vor dir verhawt es sey vor
der lincken oder rechten seiten so haw im frolich nach
zu der bloss ferrer den auch hawt und wendt dir snidt
an das schwertt so merck als bald ein schwert an das
ander glitzt so schneid im nach dem hals oder ...
... und nym den schnidt

Das ist der text von dem uberlauffen

Wer vnden remet und uberlaufft der wirt oben geschent
wann es glitzt oben so sterck dich das bestet ich zu loben dein
arbait mach waich oder hert und durch das zwifach
Glos merck das ist wen der ainer unden zu vicht wie
du im solt uberlauffen das ramen also wen du mit dem
zu fechten kumst hawt er dir den unden zu des versetz
den nicht sunder merck wen sein vnderhaw gegen dir
gatt so haw im von deiner rechten achseln oben lang ein
und schaub im den ort ~~ein~~ zu dem gesicht oder brust und
setz im an so mus er dich unden mit ... lassen fertt
er den von unden auff ... so bleib mit der lengk
Schneid starck auff dem schwertt und arbait behend
isleich zu der nachsten bloss

Item merck wen du im starck auff sein schwertt
gebunden hast schlecht er den auss der versetzung umb
dir zu der andern seiten so bind im oben mit der

Langen schneid sterckh auff der schwech seines schwerts
oben ein zu dem kopff und arwait den zu der bloss
also widerumb treib zu baiden seitten

das ist der text wie du stich und haw
abschneiden solt

Lere abschniden stich kunstlichen leren wer auf dich
Sticht den ort dick und seiner brust von baiden
Seitten duppff alle mal wiltu schreitten Glos merckh
die abschneiden treib also wen du mit dem zuvorderisten kunst
stellest sich der gegen dir als er well stechen so setz den lincken
fus vor und setze gegen im in der hut des pflugs zu
deiner rechten seitten und gib dich bloss mit dem lincken
sticht er dir den zu der bloss so wind mit dem schwert
auff dein lincke seitten gegen seinem stich die kurtz schneid
an sein schwert und setz du damit ab das dein ort alweg
bleib und gegen im sten und schreit damit zu mit dem
rechten fuss und stich im zu dem gesicht oder brust

Item ein ander stuckh wen du stest zu deiner rechten seitten
in dem pflug sticht er dir den zu deiner lincken seitten
oben ein zu der bloss so var auff mit dem schwert
und wind damit auff der lincken seitten gegen seinem
stich vor dem in den ochsen und dritt hinzu mit dem
rechten fus und stich im zu dem gesicht oder brust und
die stuckh treib von der lincken seitten auch also aus
dem pflug als von der rechten seitten

das ist der text wie du durchwechseln solt

Durchwechsel lere von baiden seitten stich mit sterck
wer auff dich binder durchwechsel in schier
findet Glos merckh der durchwechsel ist vil
und mancherlay die magstu treiben aus allen
hewen gegen den fechtern die darwider versetzen

vnd die dv hawes zu dem vnd nitt zu der blössen des
leibs die solt gar wol kennen treiben mit vnschu-
stigkait das man dir icht ansech oder sunst ein kum
die weil du durchwechselst Item das durchwechseln
treib also wan du mit dem zu vechten kumst so haw
im oben lanck ein zu dem kopff haut er dan vnd
versetzt dir zu dem schwertt vnd nit zu dem leib so las
dein ortt mit dem haw vnden durch wechseln das
er dir an das schwertt bindt vnd stich im zu der
andern seitten wirt er des stichs gewar vnd gett mit
dem schwertt dem stich bald nach mit versatzung
so wechsel aber durch zu der andern seitten vnd treib
also wan er dir mit versatzung nach dem schwertt
gett vnd das treib zu baiden seitten

Item ein anders wan du mit dem zu vechten kumst so
setz den linken fus vor vnd haltt im den langen ortt
gegen dem gesicht hawt er dir dan von oben oder von
vnden zu dem schwertt vnd wil dir das schwertt
schlagen oder starckh einbinden so las den ortt vndersich
sincken vnd stich im zu der andern seitten vnd das
treib gegen allen hewen die man dir zu dem schwertt
hawt vnd nit zu der blössen des leibs Item das starckh
merckh gar eben wan er dir vor versetzt vnd las
den ortt lassen gehen

Item hie hebt der text an vnd von dem zettel des
langen schwerts wie man sich halten sol im
schwertt alles dritt vnd mas vnd stich vnd hew
letzen straich mit stossen vnd auch alle blössen
vnd wan einer waich ist so bis du starckh vnd
wan einer starckh ist so bis du waich so wird
es in schwerch vnd starckh mit ein ander gutt in
der hutt

Item es ist zewissen das die weche die kunst bricht / und die zwai hengen und das verschieben und / die sche[illegible] und die gulden kunst die / funff stuckh die brechen den zettel auch und / ander [illegible] ain der nu bricht mit ain oder / zwayen stuckhen den es bricht ein schar der / [illegible] und ein stuckh das ander und / ein stuckh bricht den andern alles merckh

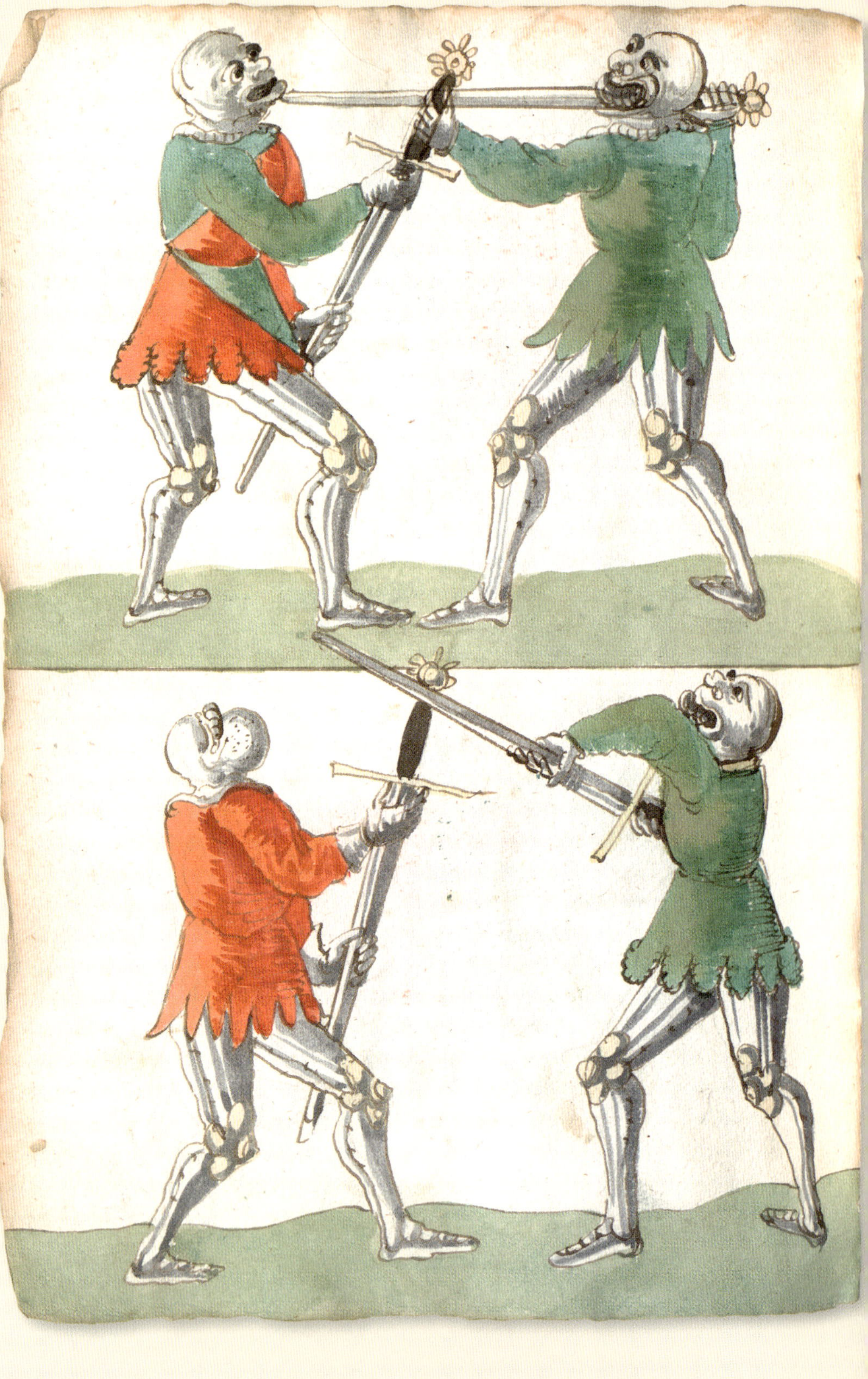

1 7 5 2 2 P 59

Hie stond drey Sön ain Reicher ain starker ain Schwacher

f. Cgm 3712 / 156

Wan hatt der Reich
weder kunst noch
weiß so vertut er
hab und gutt und
ist ein gar nichts
Reich

Wan hat der starck
kunst noch mutt
So verkumpt er
weder hab noch gut
war zu ist im sein
Sterck gutt

Wer hatt der schwach
kunst und weiß die mag
er brauchen mit guter
Sinn damit mag er
Gross er und gutt
gewinnen

Hilff herr dú ewige
wort hilff hie und

Hie tritt der maister sein herren hin zu die Schrancken
wie er sich halten sol

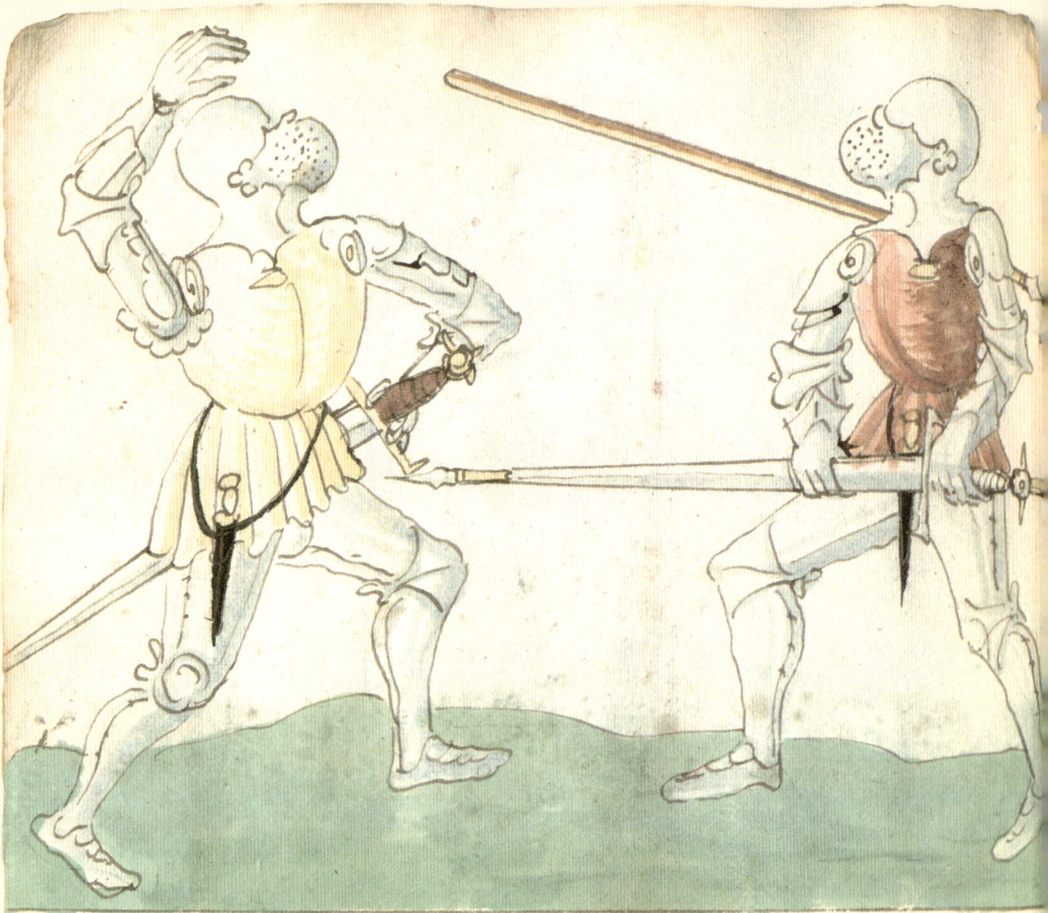

Hie komend die herrn in die schrancken zů samen
vnd wie sy den kampff volenden wöllen

Das ist die erst hůt und stand

Item So du ainen zů kampff leren wilt So merck wie
er geschickt sy oder ain gůt hertz hab das er in vertrů zů
fechten und ob er gerecht sy oder nit ist er gerecht So weiss
in in die recht kunst leg im an ain schwaren harnasch an
damit er brauchsam werd zum ersten sol er ainen harnasch
haben das er in brauchen mag wiltu ain kampff gehabt oder
gefecht zů den drien weren mit dem spiess oder schwert
und dolch So leren in des zum ersten mallen die zwen stand
und hůt wie oben gemalt statt das er in stand griff etc

Das ander Stuck

Item die ander zwen stend oder hutt wie vor geschriben sta[t]
merck gar eben das du in deiner arwait gwiß stust
und wolbedacht wie du in den man verlest zuwarten
und sick dich wol an in wie oben gemelt statt

Das dritt Stuck

Item vermerck den dritten stand der haist die sterckst hutt so
einer den knopff gegen die brust und hatt das schwertt auff
der linckhen seitten So truck dich wol mit deinem
ort gegen der brust gen im weit aber enhalt fest schau
das du dich nit verfuren lassest dann es gett vil fechte
arwait darauß

Das vierd haist die haftung

Item vermerck so ainer auff dich arbait als vorgeschriben
stat so schaw du gar eben das du merckst kunst auff die
hafftung oder ansehen auff seinen fuß welchen er vor setzt
trifft du in so fal mit deiner brust auff das schwert und
druck es wol nider so stichst du in wol nider durch den
fuß wie oben gemalt ist das ist das erst ansehen oder arbait
in dem kampff

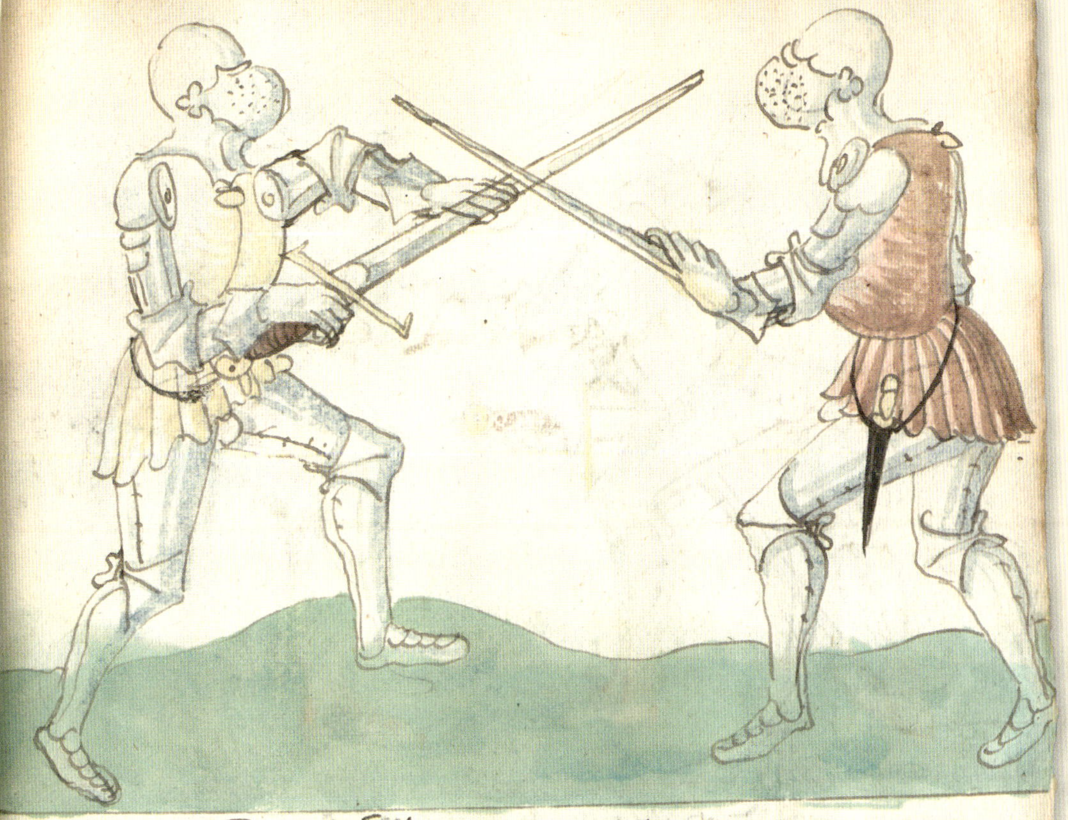

Das stuck haist anworffen

Item So du sichst das dir ainer wil ansehen nach dem fuss
wie vor gemalt statt So fe auff mit deinem Schwertt
Stoss im nach dem gsicht und tritt mit dem andern fuss
ein So prichst du im die hut und wie oben gemalt statt
und zech dich wol

Das Sechst ist die hafstuck

Item Wer aine umb den obern stich wie oben gemalt ist
entwichsch gaben will und sticht dir nach dem hindern fuss so
sol mit deinem schwert mit deinen baiden henden auff
sein schwert trucke und nider so machst du dich ledig
wie hernach geschriben statt und bist gewiss in der arwait
wie oben gemalt statt

Das sibent haist der mordschlag

Item die dritt haftung oder anrechnung So dü das gwar
wirst So nim das schwert bey dem ort und ker den knopff
vür schlach im zu dem helm So wirtt er tenmisch in sein
helm wie oben gmalt Stett So wist dü ledig

Das acht haist haftund

Item So dir ainer nach dem kopff sticht oder schlagen wil So farh auff mit deinem schwert und setz im an mit dem ort an dem tenckhen elenpogen damit hast du im den mordschlag brochen und er wel in dich und schreib in die sterck wie oben gemalt stett

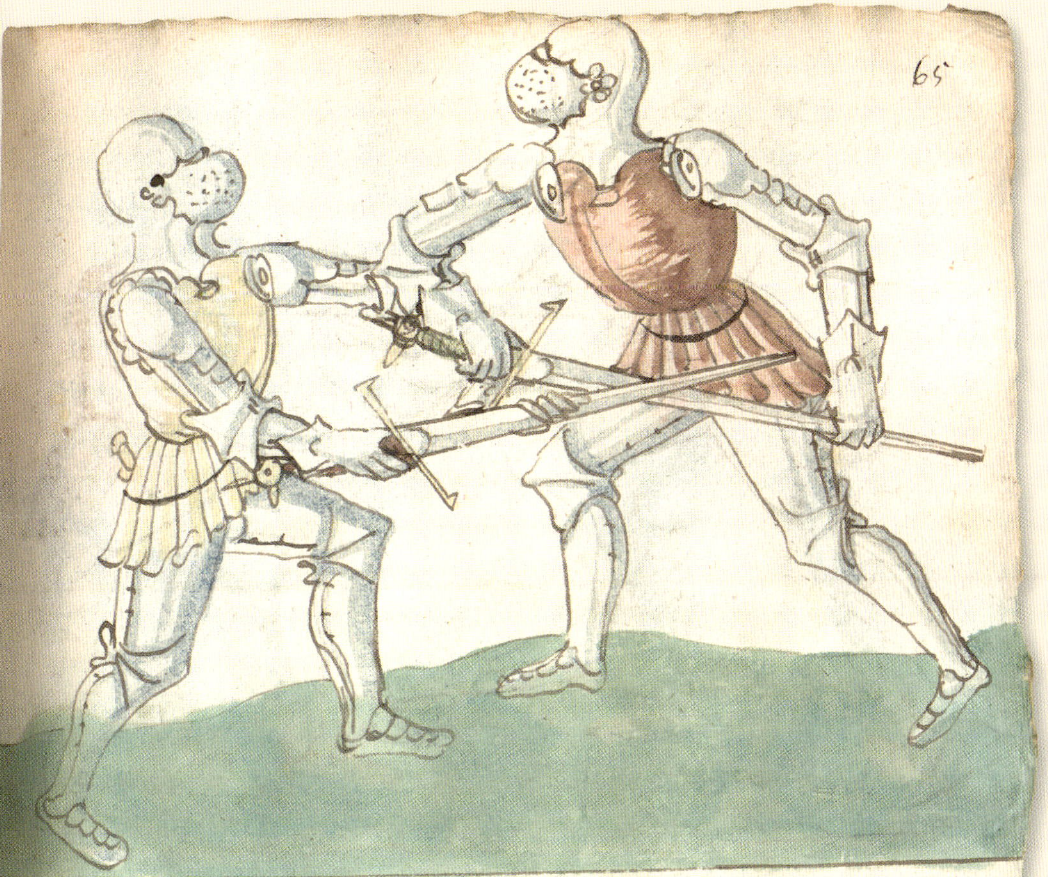

Das wůndt uberwinden

Item vermerck. die wůndt erwaist So dir ainer an gschoß hast und wil dich schiben wol zu ruck So nim dein Schwert bei dem ghiltz in die tenckhen handt und drit mit deinem rechten fůss vir und wind im mit dem knopff noch dem gsicht So prychst du im das an Gschoss und hast in wie vor gmelt statt und der arwait der buch die darauff steet

Das zehent Stuckh

Item So du sichst das dir ainer nach dem gsicht arbait So drit vir mit dainem rechten fuss vnd mit deinem Schwertt wol in die höch vnd stoss mit dem schwertt innerhalb seins schwertts vber sein rechten arm das hast zwischen baiden armen durch gestochen vnd druck wol nider drit mit dem tenckhen fuss für so hast du das ansehen oder hastuns auff seinem rechten fuss

Das ailft Stuck

Item So dů sichst das dir ainer zwischen baiden armen
ein durch wil stossen und dir die sterck hat abgewunen
So drit im mit dem rechten fůß und wol auff mit
dem ort nach seinem gsicht wie oben gmalt stet So
hast du in geprachen

Das zwelfft Stuck

Item So du sichst das dir ainer nach dem gsicht arbait
So drit dir mit dem rechten fuss vnd wind mit deinem
knopff mit deiner rechten hand vnder seiner tengken
mit dem schwertt zeuch wol an dich wie du oben vor dir
gemalet sichst

Das dreitzehenst Stuck

Item So dü sichst das dir ainer durch windt wie vor
gemalt und geschriben Statt So trit mit deinem rechten
füss für und begreiff Sein schwertt mit sampt deinem
in dein linckhe hand und wind vast nider das haist der
klemer wie oben gemalt Statt

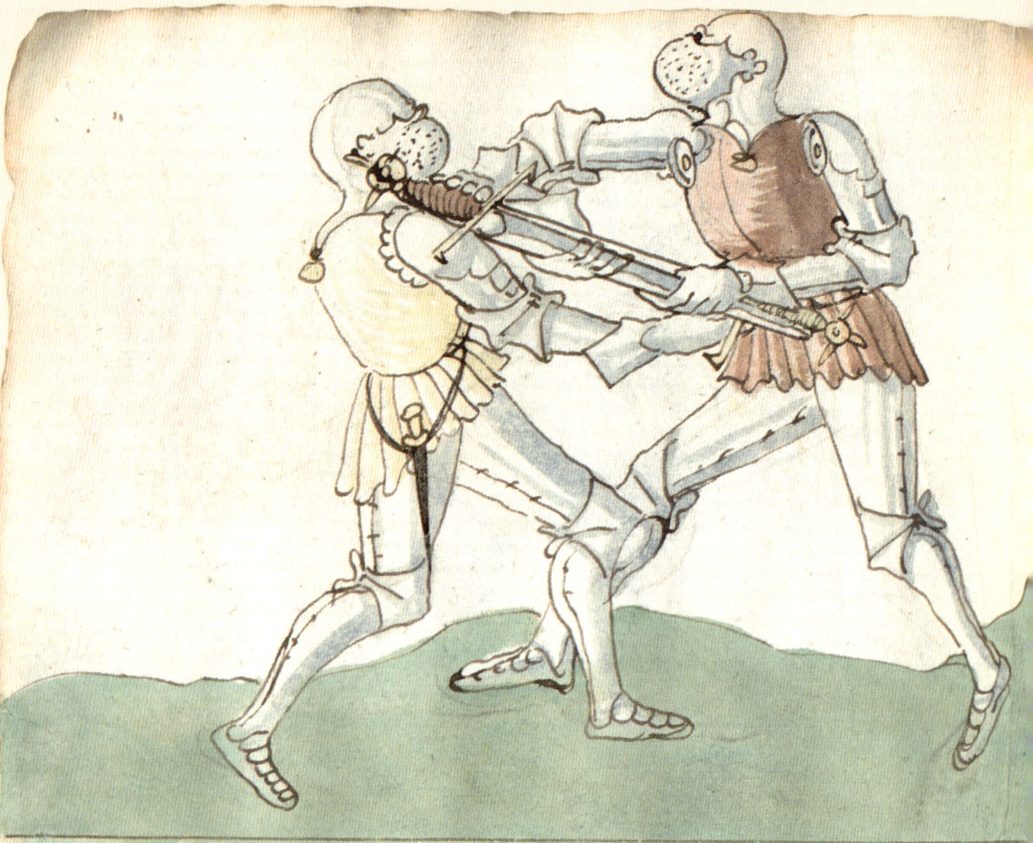

Das vierzehenst Stuck

Item So du sichst das dir ainer wil fallen in den klame
So drit mit deinem rechten fuß vir wind mit dein schw-
ert und knopff wol auff das du im an den hals kumst
zeuch wol zu dir So wirfstu in ueber den fuß wie
oben gemalt Stet

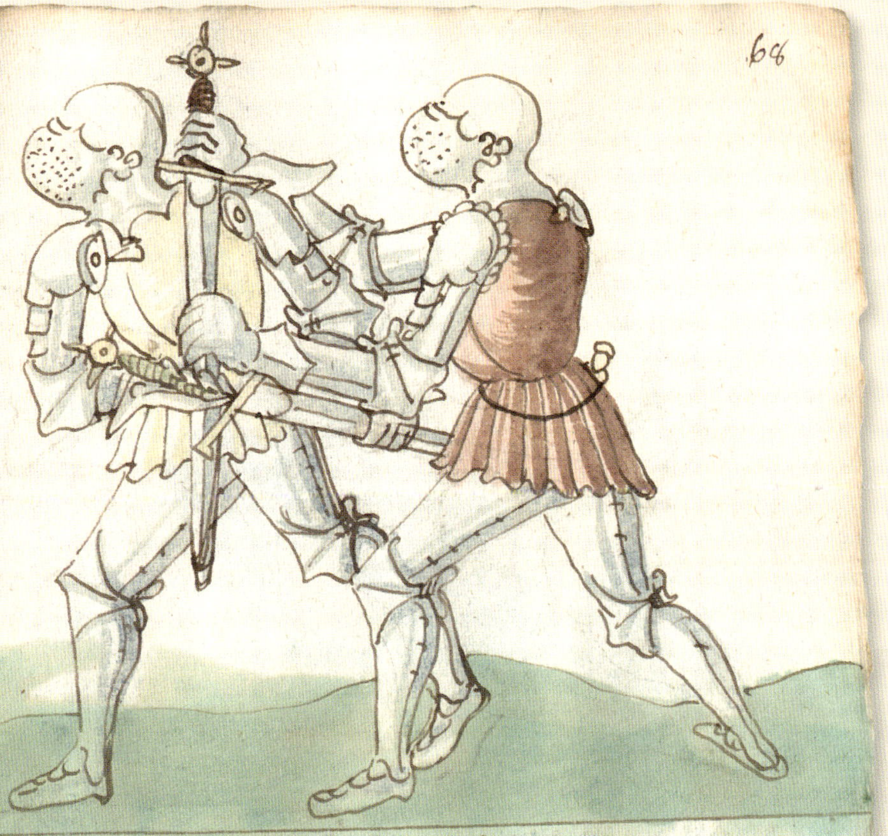

Das funffzehent Stuck

Item So dir ainer nach der vnder hastnuss oder aufsetzen oder
will dich ~~[illegible]~~ oder die pain werffen So thu im also
gantz auff mit dem schwertt vnd dut wol die hinder
dein füss vnd fall im mit dem schwertt auff sein schwert
vnd truck jms wol nider auff den hals wie oben gemalt
Stat das ist der dritt haupt pruchanner

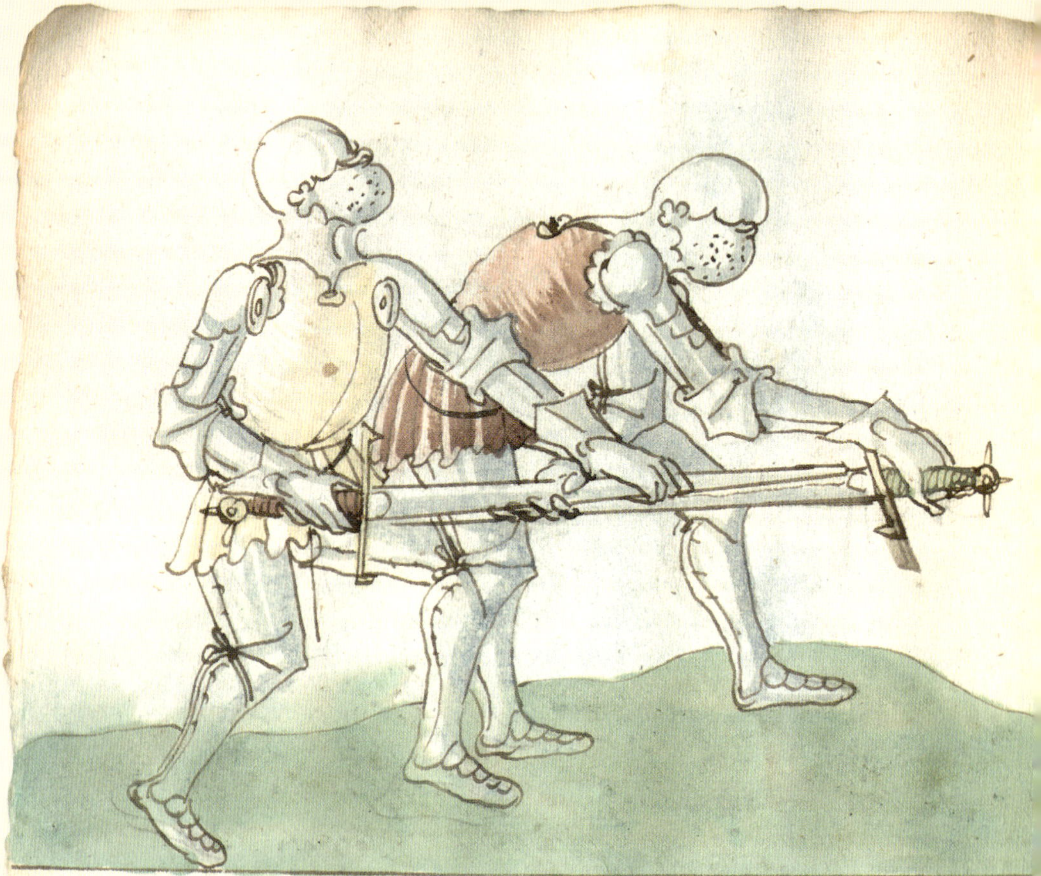

Das Geschrenckt Stück

Item So dir einer umb den halß vallt vnd wil dich werff[en]
So zuck den fuß an dich vnd tritt wider ein mit deinem
rechten hinder seinen rechten vnd wind mit deinem —
lincken wol vnder sein lincken elnpogen hinder sein ve[r]-
schisstu druck von dir So wirffstu in auff das angesicht
mit aller gwaltt starck das ist der ander haubtbruch

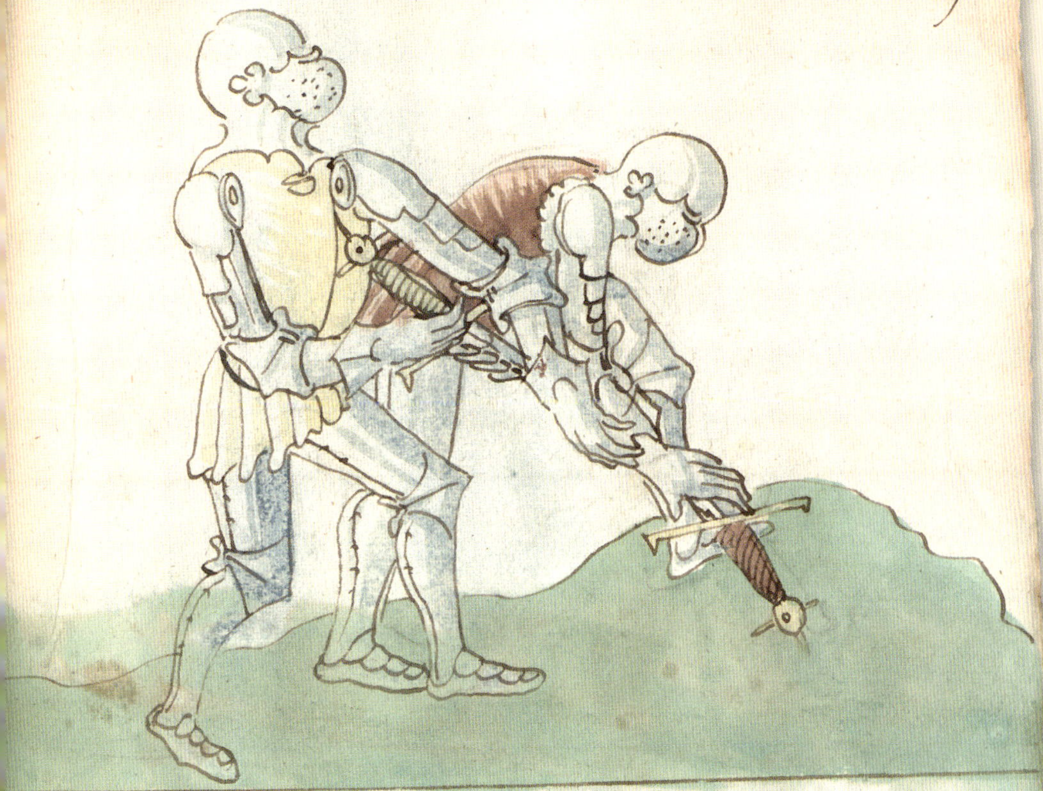

Das Sibenzehenst Stuck

Item So du sichst vnd empfindt das dich ainer also verkeück
en will So dritt mit deinem rechten fuss für ausserhalb
sein rechten nim dein lenncke hand vnd greiff an sein
rechten elenpogen Stoss wol von dir So hast du im das
vorder Stuck gebrochen wie oben gmalt Statt

Das achtzehenst Stuck

Item So dich ainer hinweg stost mit dem linckn dein
rechten ellenpogen So var auff mit deinem rechten arm
und mit dem lincken arm lenck bey dem ellenpogen
mit deiner rechten handt umb sein halss nim in wol
bey seinem part oder visier zeuch in wol in die höch
oder zück dich nider welches du wilt das ist vast ain
guetter prauch und stuck wie oben gemalt stat So
magst du dir ler wol behalten

Das newnzehenst Stuck

Item So du sichst das dir ainer vmb den halss wil fallen vnd wil dich mit seiner rechten dein tencken ellenpogen zuruck stossen So trit mit deinem rechten fuss für hinder sein knie mit deiner lincken handt nimm zwischen seine pain in der kniepieg heb wol in die höch stein rechten fuss So wirfstu in an den rucken wie oben gemalt.

Stark

Das zwainczigst Stuck

Item So du sichst das dich ainer also vassen wil wie oben
gmalt stat So greiff mit deiner rechten oben uber sein
halß in sein rechten ellenpogen zeuch sein tencken arm
wol an dich und legk dich wol auff in So pringst du
in das ander stuck und druck in nider zu der erd
wie oben gmalt stat das ist vast ain guter
pruch

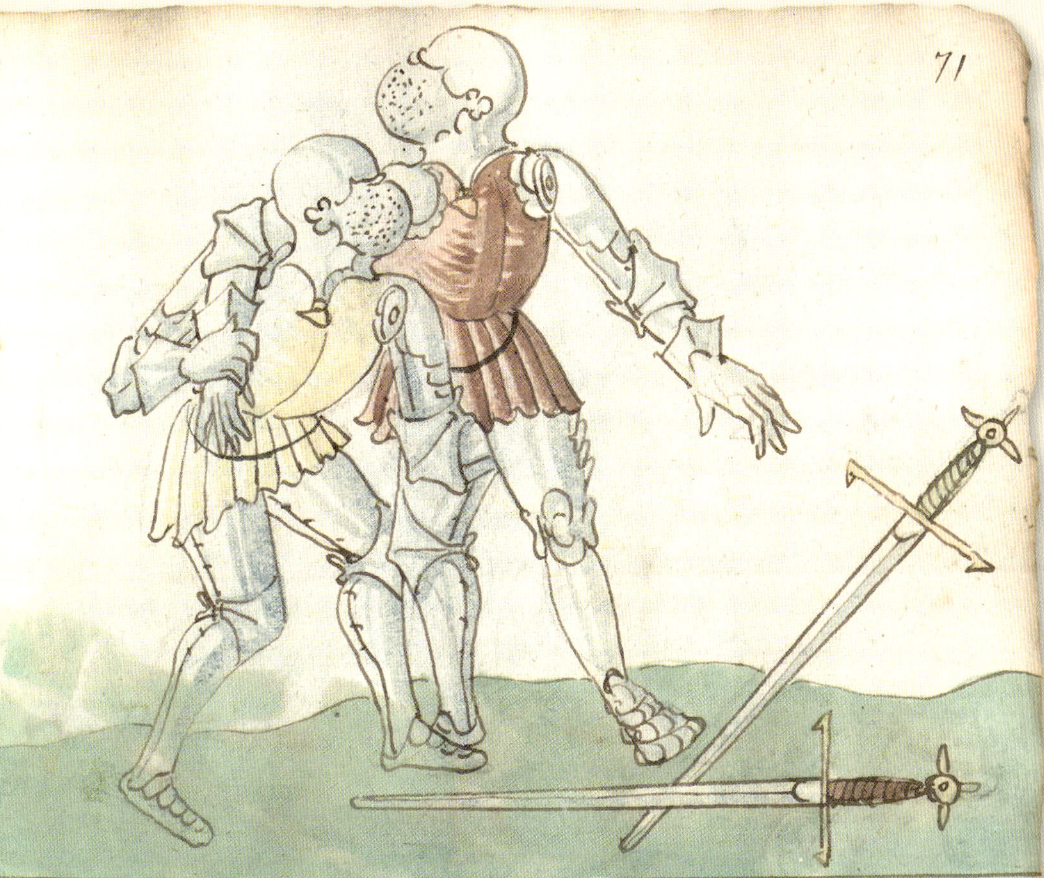

Das ainundtzwaintzigist Stuck

Item So du Sichst oder empfindst das er dich also zu der erd
trucken wil So greiff mit deiner rechten in Sein rechts
glenck vornen bej der hand vnd heb mit deiner rechten
Sein rechte hand auff oder Sein halss vnd trit mit deinem
tenncken fuss da ein vnd greiff mit deiner tenncken hand hinder
in Sein harnasch oder gesess oder zwischen der pain welches
dir werden mag heb wol in die hoch So wurfst du in
an den rucken wie oben gemalt Statt vnd ist vast
ain gut Stuck

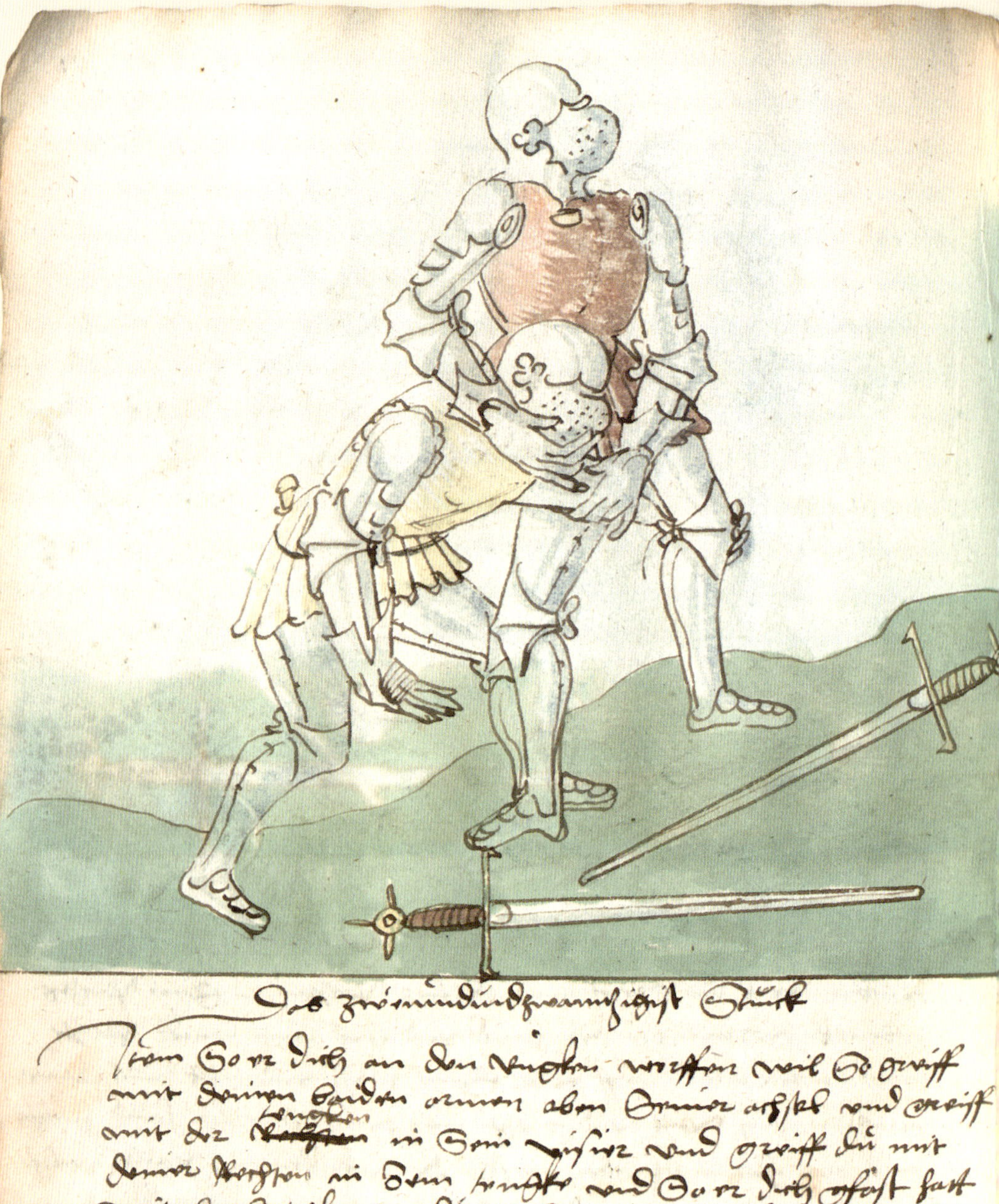

Das zweiundzwainzigist Stuck

Item So er dich an den Ringen wessen will So greiff mit deinen baiden armen aber seiner achsel und greiff mit der ~~rechten~~ lincken in sein visir und greiff du mit deiner rechten in dem rucke und So er dich gfast hat mit der lincken hand bey deinem rechten paini und wil dich heben zum wurff So leg dich walauff in ruck in zu der erd wie aber gemalt Statt So hastu im sein arwait gbrochen

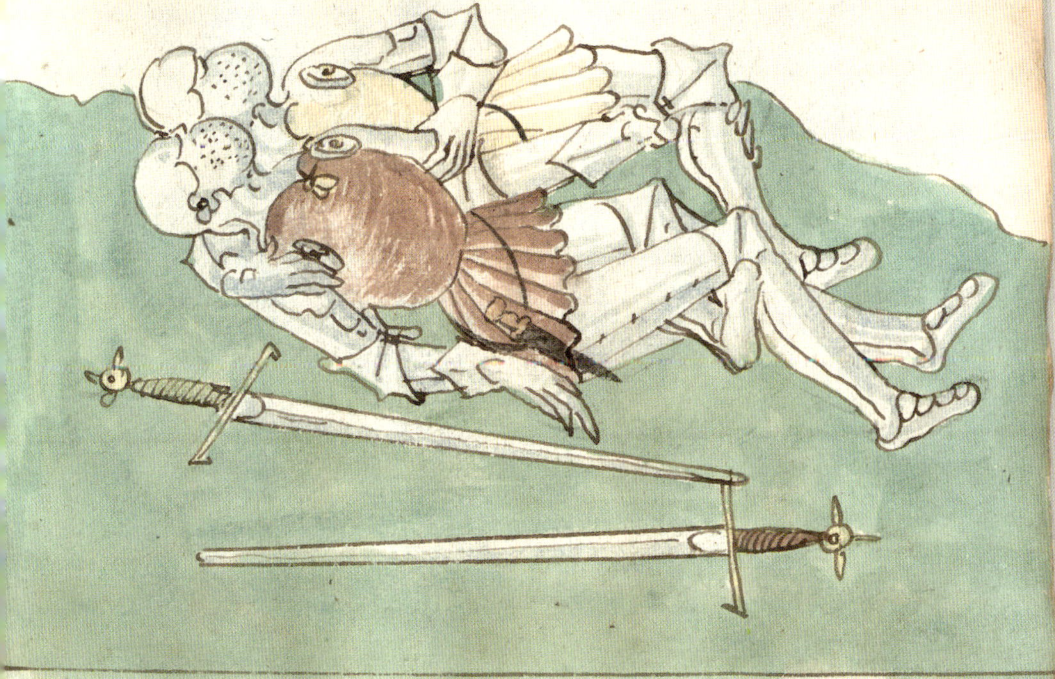

Daz drÿundzwaintzigost Stuck

Item So dich ainer alsz nider truckt zu der erd wie vor gmalt Stat so fal flucks hindersich auff den Rucken So musz er mit dir fallen So wirst du ledig von im und mit laufs spring bald auff und stosz mit dein rechte knie an sein rechte fusz So musz er von dir lassen wie oben gmaltt Statt

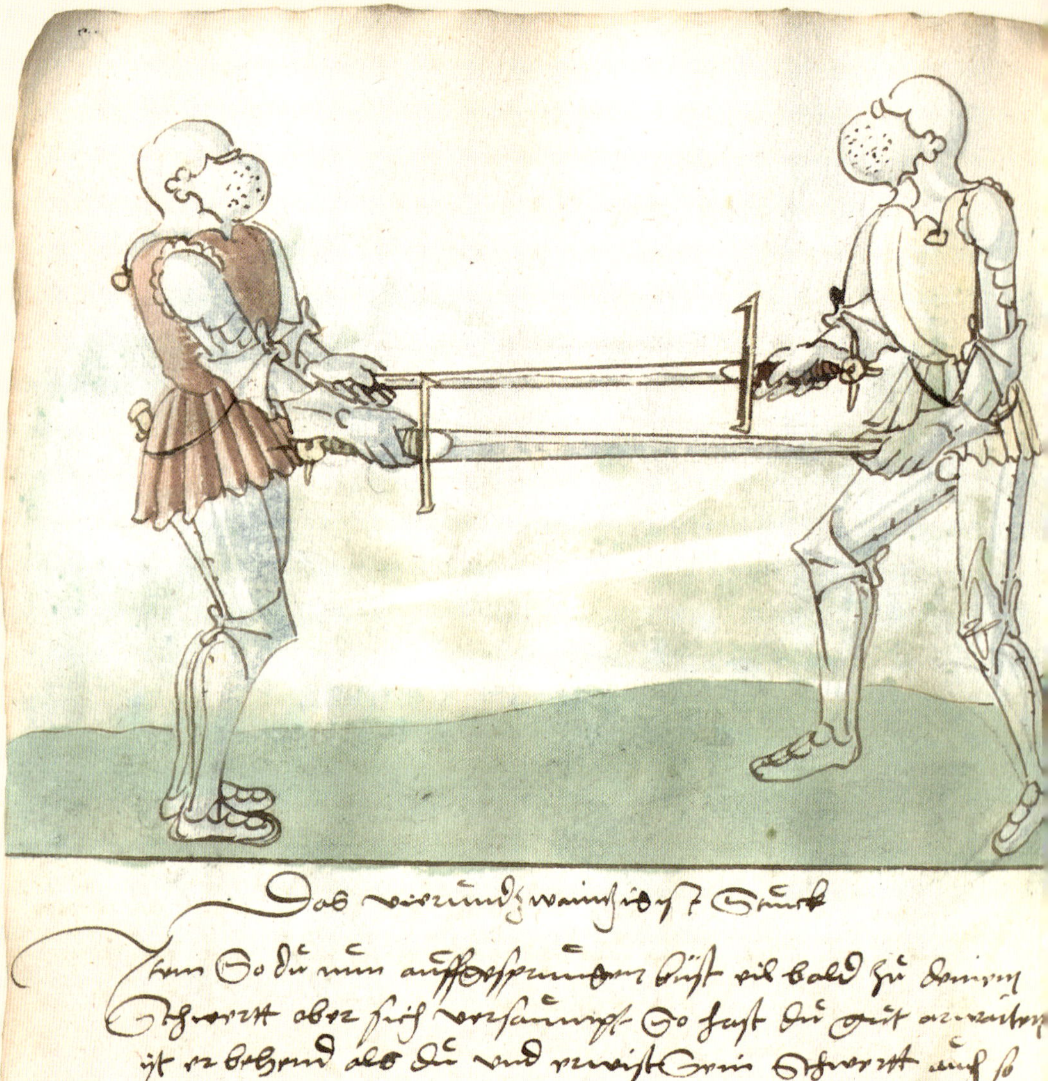

Das vierundzwainzigest Stuck

Item So dü nun auffgesprungen bist eil bald zü deinem
Schwertt oder sich versaümpst So hast dü gut erwarten
ist er behend als dü und zeucht sein Schwertt aüch so
eil mit deinem langen Stich aüff in und begreifft mit
deiner tencken handt sein ortt begreifft er das dein aü
krieg dem ortt So ster fest schaü ob er starck oder schwach
sey darnach waist dü zü erwarten wie aber gewalt
statt das haist der anfang der schwer

Das fünffundzwainzigist stuck

Item So du empfindst das baid Schwertt ghanngen sein wie vor stett So wind flucks durch mit dein baid armen kreucz weiß ober ain ander mach auff dein achsel buck dich nider vnd zeuch wol an dich baide schwertt wie oben gemelt Stost vnd kere in den venstern das haist die Schrankhut

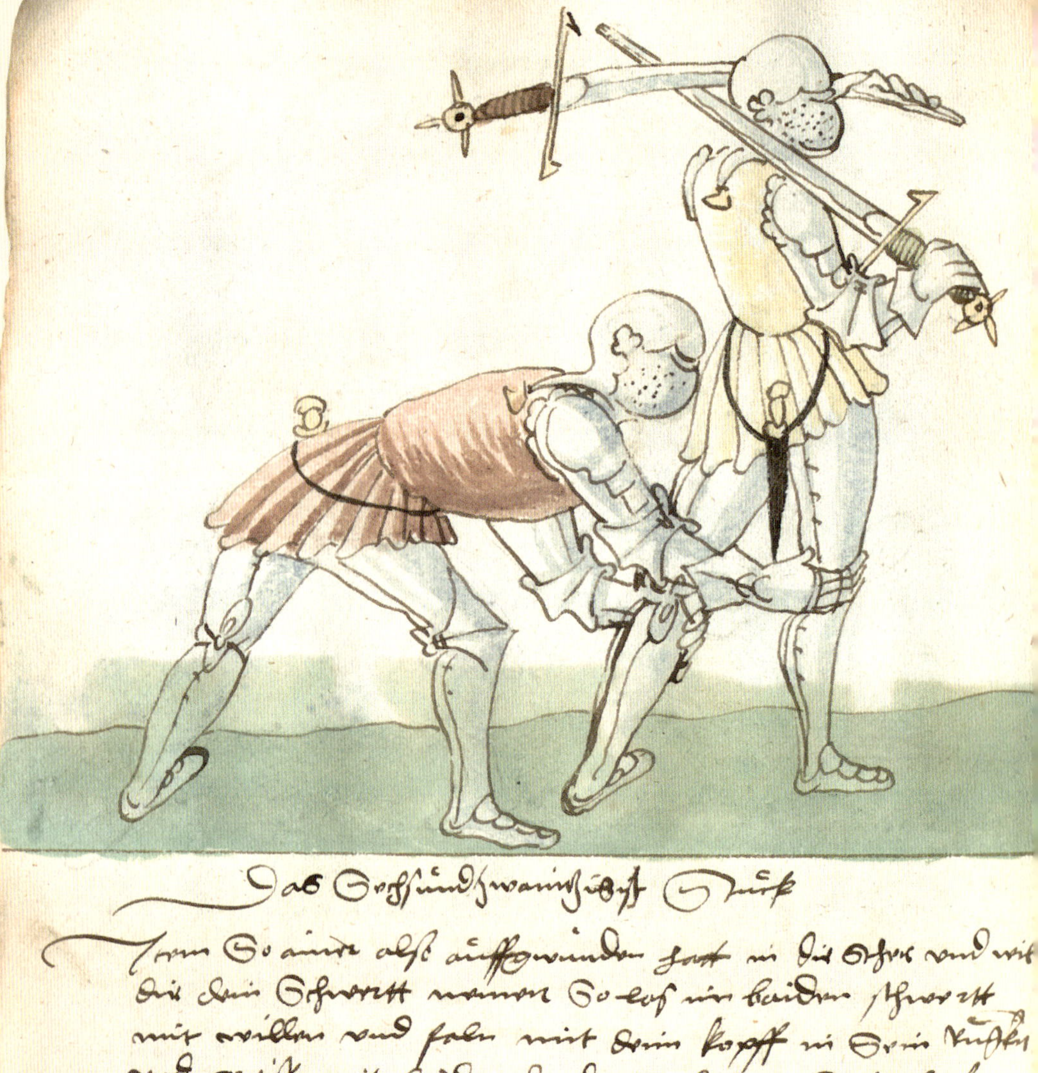

Das Sechsundzwainzigist Stuck

Wenn Er dir also auffgewunden hatt in die obern und wil
dir dein Schwertt nemen So loß im baiden schwertt
mit willen und fal mit dem kopff in sein ruckhen
und greiff mit baiden henden unden in sein füeß an
die kniepüeg zeuch wol an dich und mit dem kopff
stoß von dir So würffstu in auff den ruckh wie oben
gemalt steet

Das Sibermündheraich ist Gut

Item So dich ainer werffen wil wie vor stet So trit mit dem tencken fuess für mit dem rechten ker dich bald umb und stoss in mit dem knopff zu dem hals So du das gewar wirst und get dir auf die arwait und wil dir nach dem hals stossen mit dem knopff So vall im mit deiner tencken hand an sein rechte vornen an das schultz So brichst du im den stos greiff bald mit deiner rechten hand nach deinem swertt So kumpst du wider in die arwait

Der schiernprandt

Das achtundzwainczigist Stuck

Item So ir payde an ander kumpt in die arwait So ste vest und schaw ob er starck auff dich wel lauffen oder nit So se mit deinem schwertt vor im umb im winckel und thu als wellest du in verfieren das stuck haist der Schiernbrand und gat vil arwait darauß

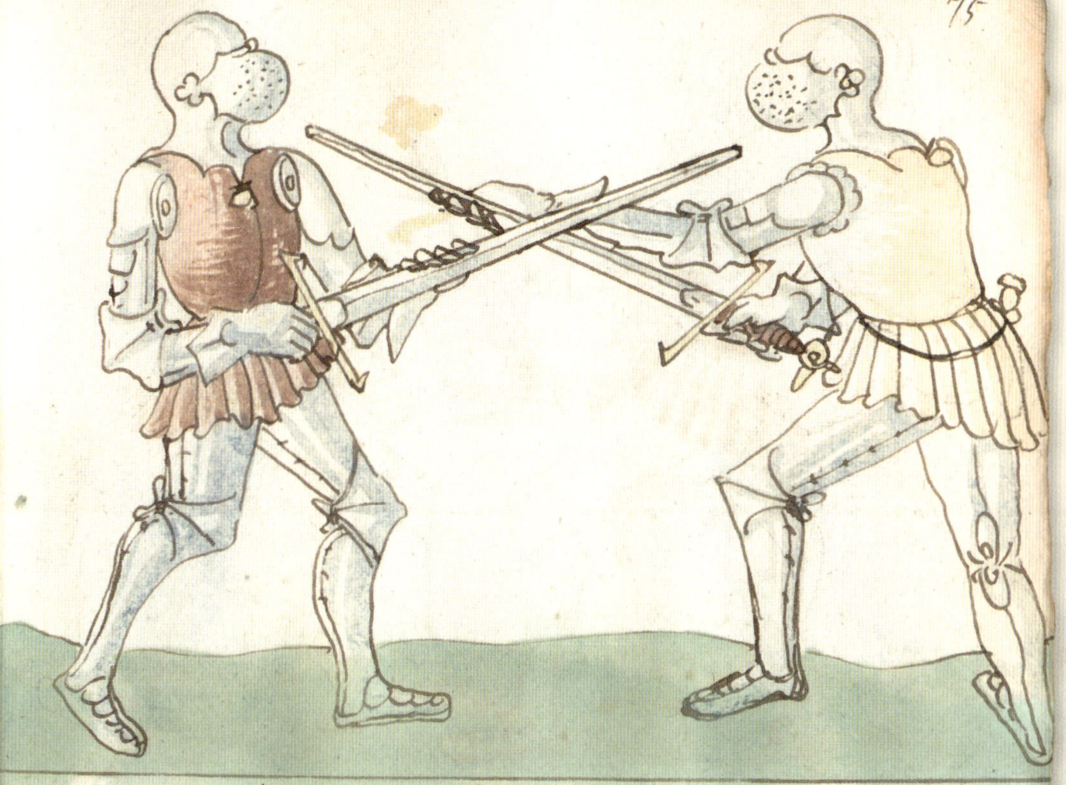

Das neunundzwainzigist Stuck

Item So einer also Starck in dich will lauffen und stost
dir oben nach dem gsicht und besorgt der hefftung oder
aufsetzen und stett vor mit seim rechten fuß und [illegible]
[illegible] auff mit deiner lincken handt an sein schwertt [illegible]
an sein rechten handt an sein schwertt wie oben gemalt
stett und merck den dritt gar wol

So drit du ein mit deim rechten fuß

Das dreyssigist Stuck

Item So ainer Starck auff dich lauff vnd wil dir dir vnder
haftung oder aufschung begreift vnd stet vor mit seiner
lincke So drit du mit deinem rechten fuß wie zu seim
lincken vnd val mit dein schwertt mit deiner lincken
hand an sein schwertt bei seiner lincken handt truck
vnder weg von dir So stost du im mit dem knopff in das
gsicht wie oben gmalt Stett ist vast ain gut Stuck

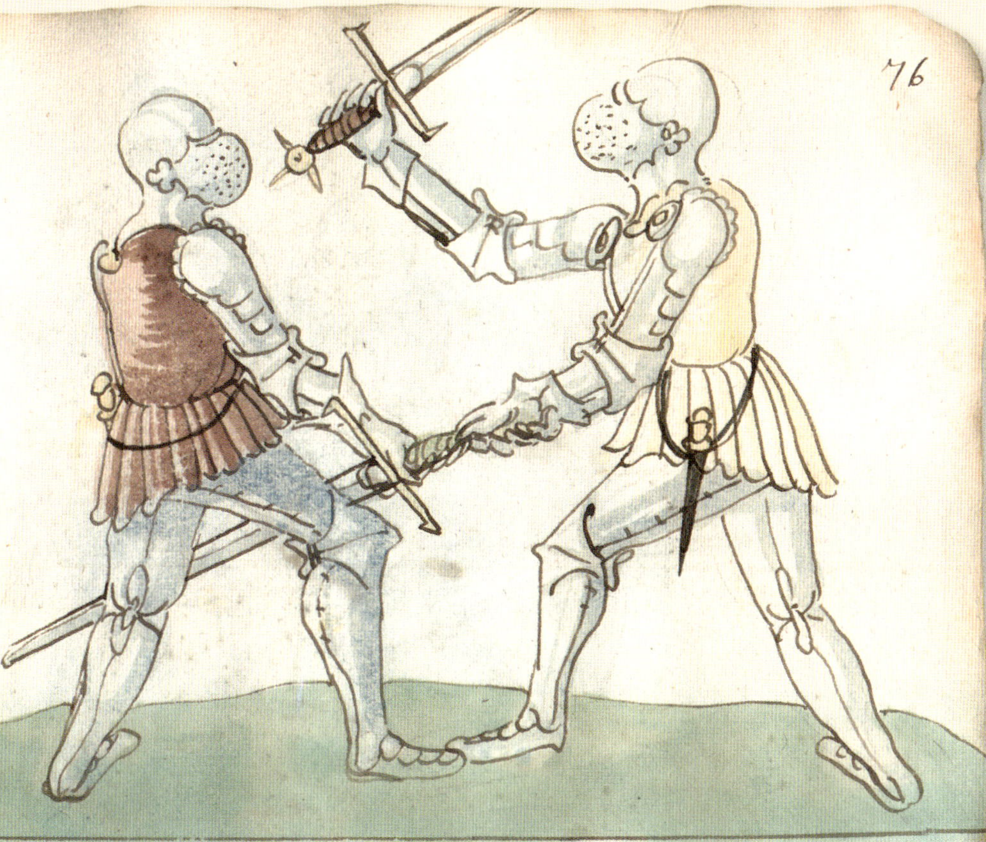

Das ainundreissigest Stuck

Item So dich ainer vbereilt vnd kanst zu kainer gewait kumen So greiff mit deiner linckh handt in sein schultr bey dem kropff vnd rechten hand in das viesier wirs aber gwalt Wert das Stuck haist die ledigung

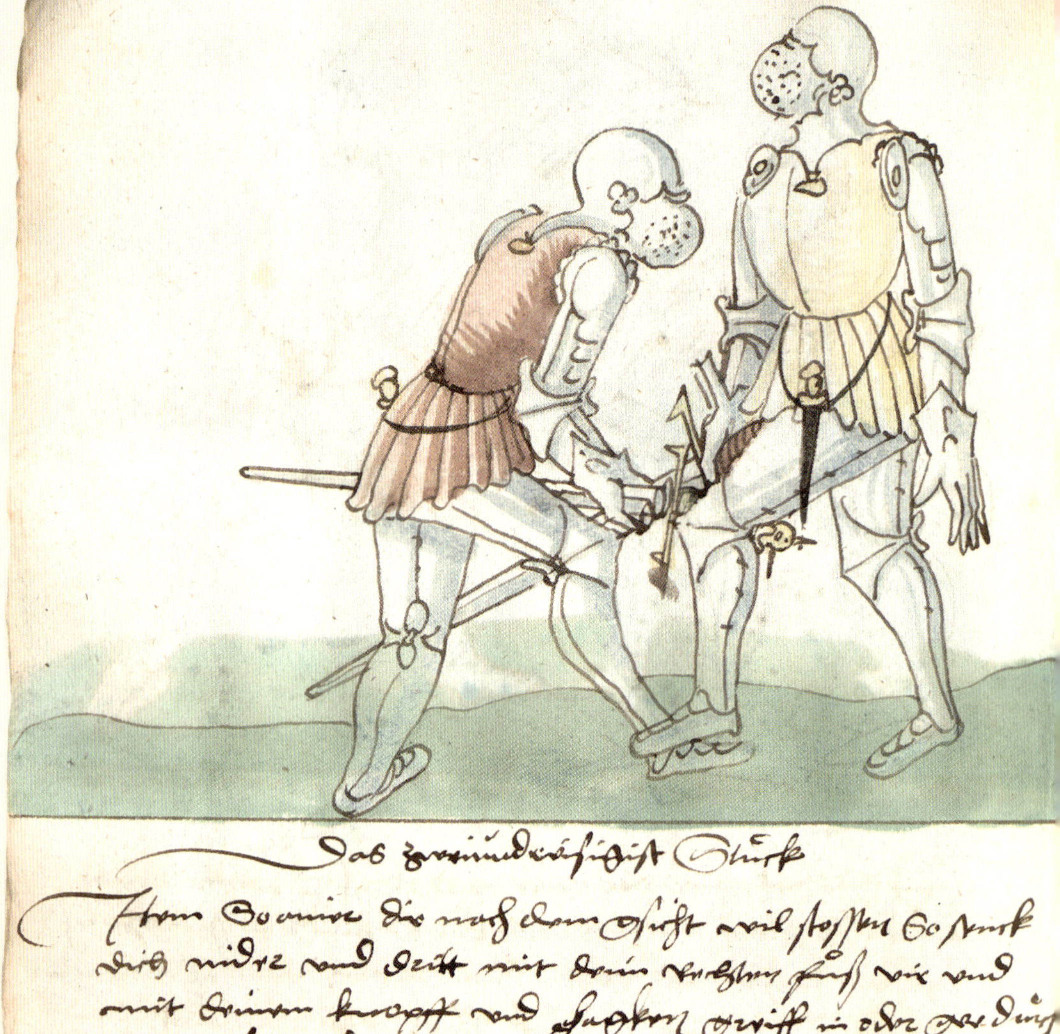

Das zwaindreissigist Stuck

Item So einer dir nach dem gsicht wil stossen So senck dich nidr und tritt mit dein rechten fůss vir und mit deinem kopff und hacken greiff im oder gedůng in sein kniepůg heb auff den fůss So wirfst du in an den rucken wie oben gmalt Stat das stůck gut von baiden Seitten oder arwait von baiden seit

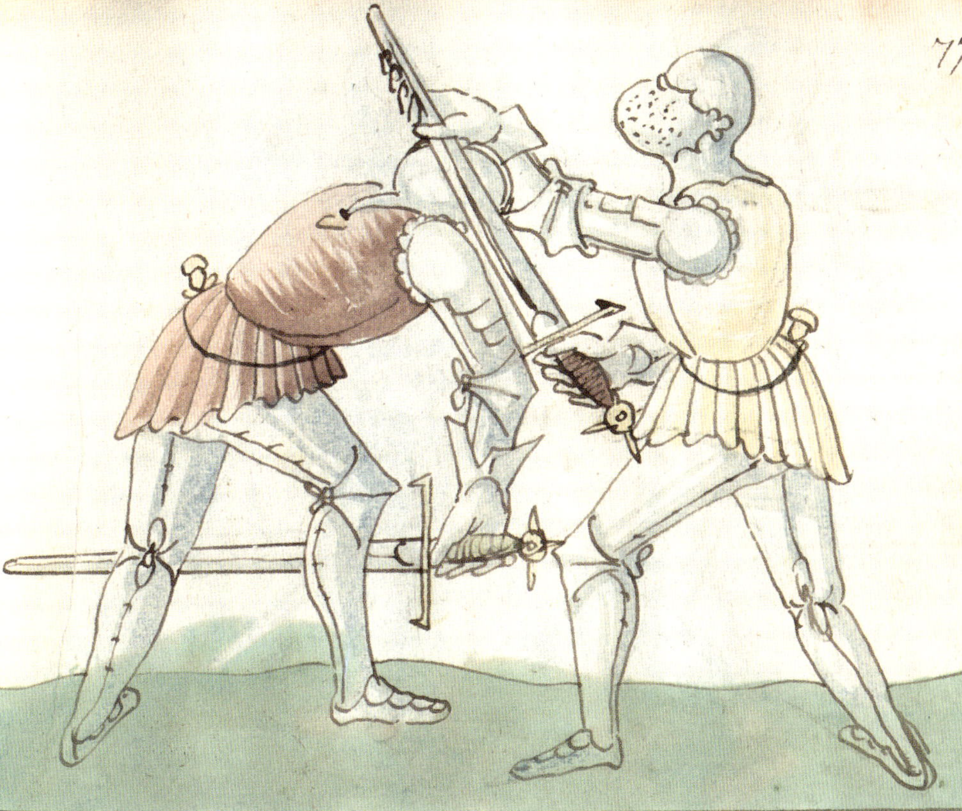

Das dreyunddreissigist Stuk

Item So sich ainer senckt nach dir vnd sett dir durch
die pain vnd wil dich werffen So zuck dein rechten
fůß an dich vnd sel im mit dein schwertt am sein
halß wie oben gemalt stett So hastu im das vorder
Stuk gesprochen vnd druck wol nider

Das vierunddreissigist Stuck

Item So dir ainer das schwertt also an den hals gelegt hatt So far vnden auff durch dein schwertt vnnd mit deinem schwertt wol auff tritt mit deinem tencken fuß ein hinder sein pain leg im das schwertt an den hals vnnd mit deinem elenpogen stoss im an sein gurtt druck wol hindersich mit ainem gwalt starck so wirfstu in vber das pain an den ruckenn

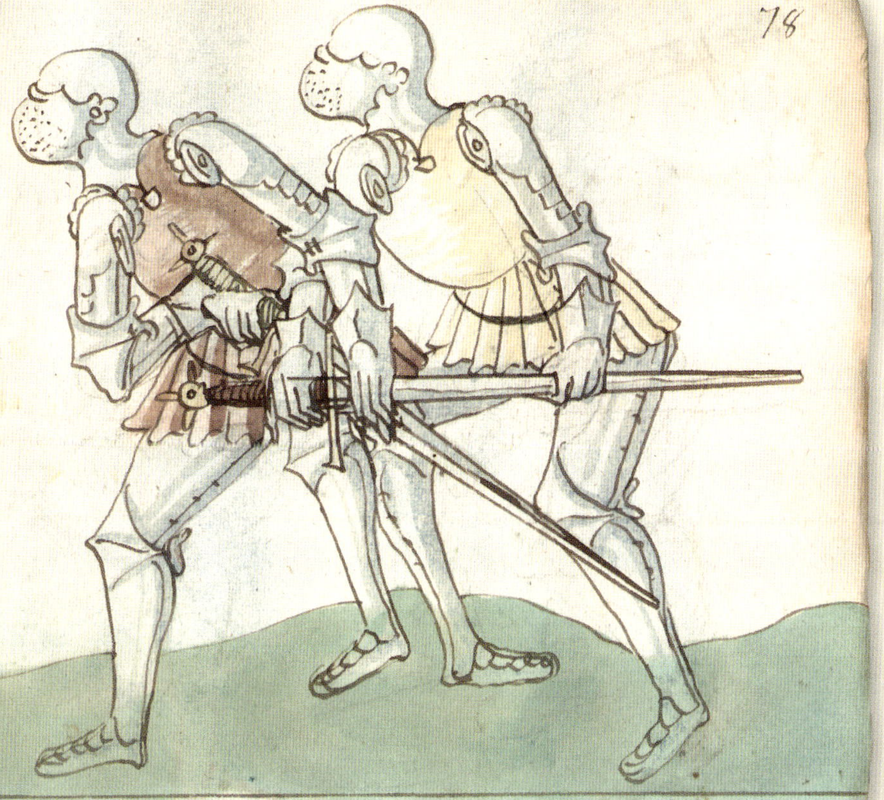

Das fünffundviertzigist Stuck

Item So du sichst und empfindst in der arwait das er dich werffen wil und dir das schwertt an den hals gelegt hatt wie vor statt So trit mit dem rechten fuss ein und wind mit dem schwertt durch sein bainden arm das der krapff und das gehultz vorster truck wol nider und Druck dich So prichst du im den wurff wie oben gemalt statt das ist die viertt arwaitt aine im kampff

Das Sechsunddreissigist Stuck

Item vermerk die fechten stend vnd gewalt So dich ainer also nidertruckt wie vor statt so ste auff mitt dem schwertt vnd lug wie du mit den fuessen ledig werdest werden das kan man also grüntlich nit schriben oder malen So du auff bist So henck mit dem schwertt vber den kopff Stoss im flugs nach der hefftung wo der helm auff ain ander gatt das haist die hefftung oder anstechen so stost du ain wund wie oben gmalt statt

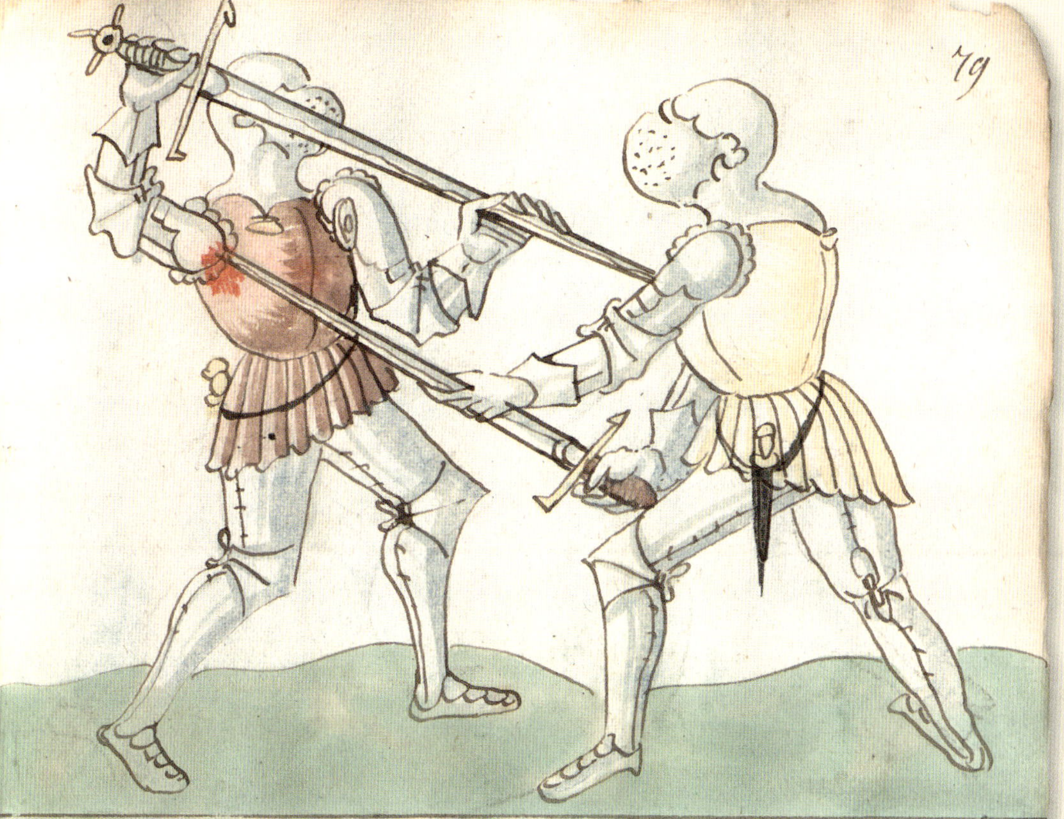

Das Sibenunddreissigist Stuck

Item Wen du sichst das dich ainer also will stossen nach der kestung an dem hals So setz dich nider und senck dich wol nider in die mass und lenck und stoss im under die vechsen yrchssen da der harnasch tailt ist So stost du im durch das panntzer blutriss wie oben gmalt ist darnach sichst du gutte arwerck

Das achtundreissigist Stuck

Item So du also sichst das dir ainer die hastung und ansetzung sucht under deiner rechten üchsen und dein stoss nit verbringen magst wie vor stett so fall behend mit deiner gewapenteten hand und schwertt auff dein linke hand truck wol nider wie oben gmolt statt ist vast ain guter bruch

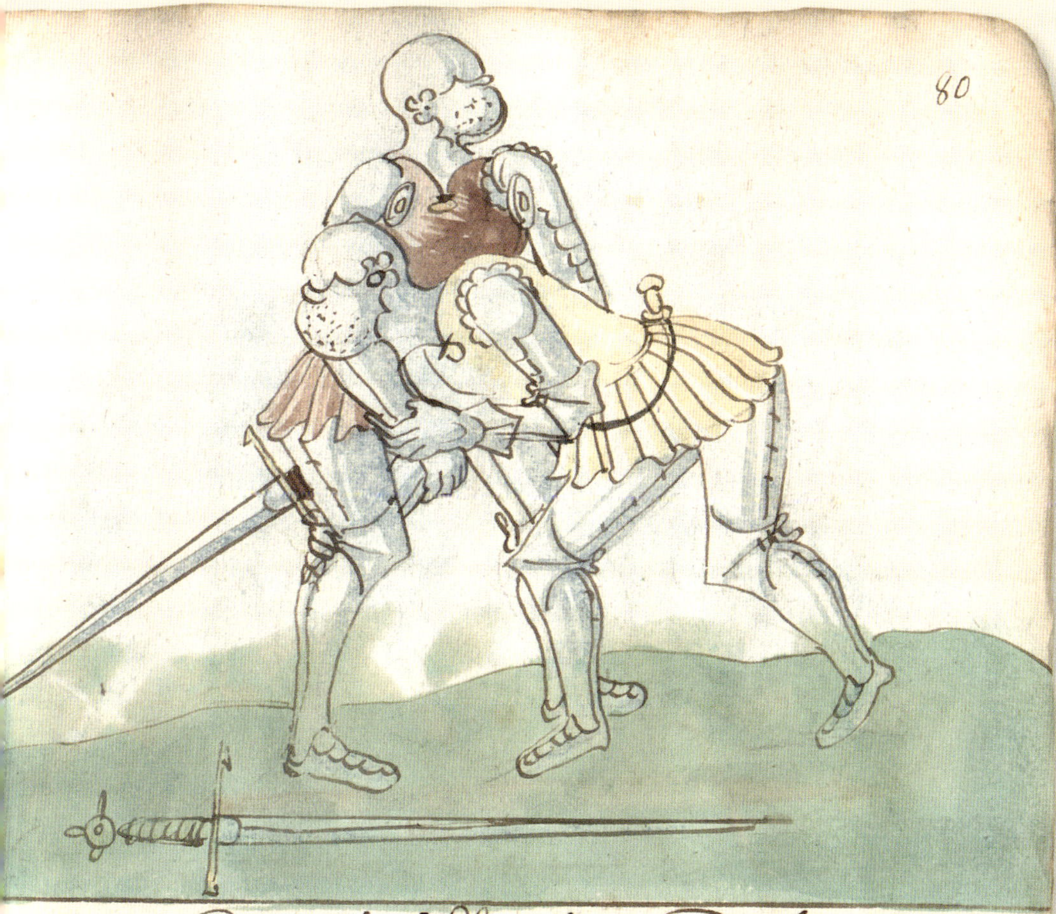

Das neünunddreissigist Stuck

Item So dir einer verfelt nach denen wie vorstatt So lass dein schwert fallen und greiff mit deiner linckhe handt auff sein rechte hand und trit im mit deinem rechten füss und greiff mit deiner rechten hand zu seinem rechten pain in die kniepüg zück dein füss an dich oder heb in auff und drück mit deiner achssel wol die So wirffstu in an den rücken wie oben gmalt Statt ist vast ain gütt stück in dem kriegen zu dem kampff von baiden Seitten gestes

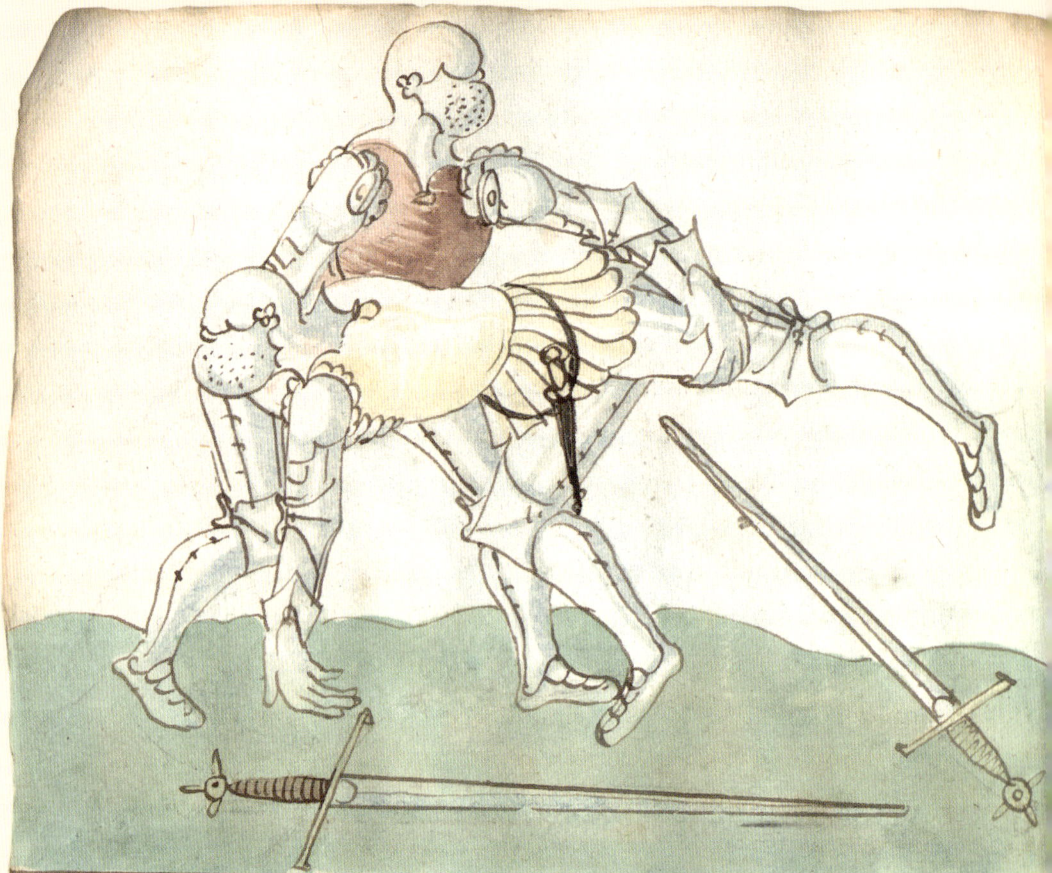

Das viertzigist Stuck

Item So dir ainer also underlaufft in dem Ringen
und wil dich werffen So mach dich ledig mit den fuessen
und greiff mit deiner rechten hand under sein gechssen
und greiff mit deiner lincken hand ober sein fuessen
in das gsess So wirfftu in an den rucken wie oben
gemalt Statt und hernach findst in den künsten
und bis gewiss in der gewalt es sey im schwert oder
im Ringen

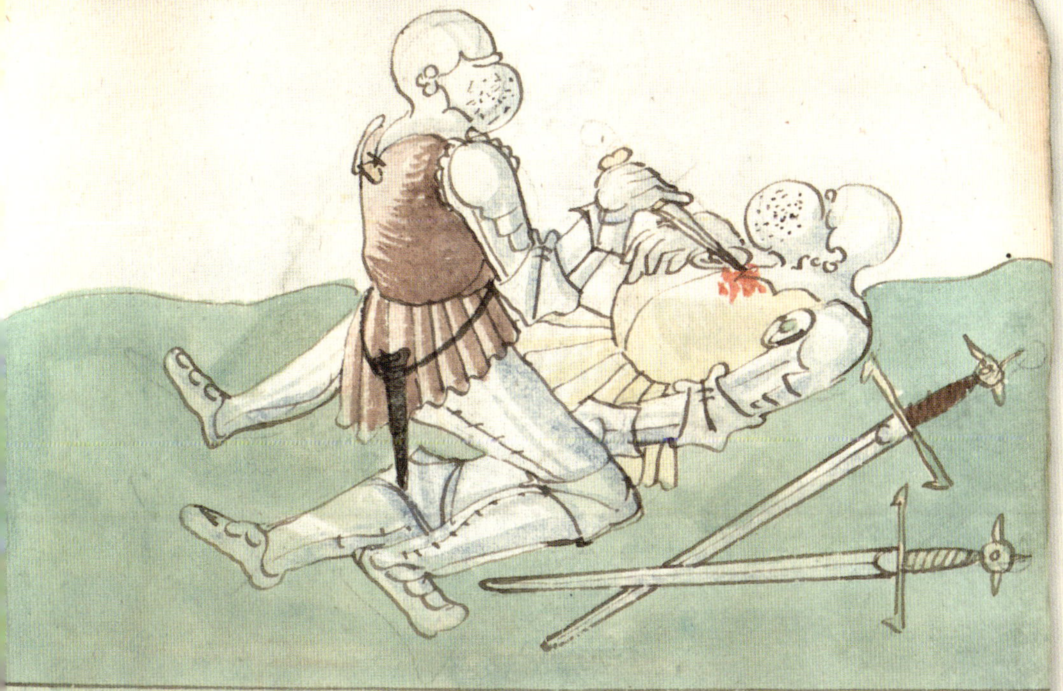

Das ainund viertzigest Stuck

Item So du ainen also auff den rucken hast geworffen wie vorstet So schau bald das du auff in kumpst vnd mit dein dolch die hastung suchest wo du sy gewinen magst So du in gewunnen hast wie oben gemalt Stat So sprich im zu ob er sich zu dem halben erab geben well oder nit

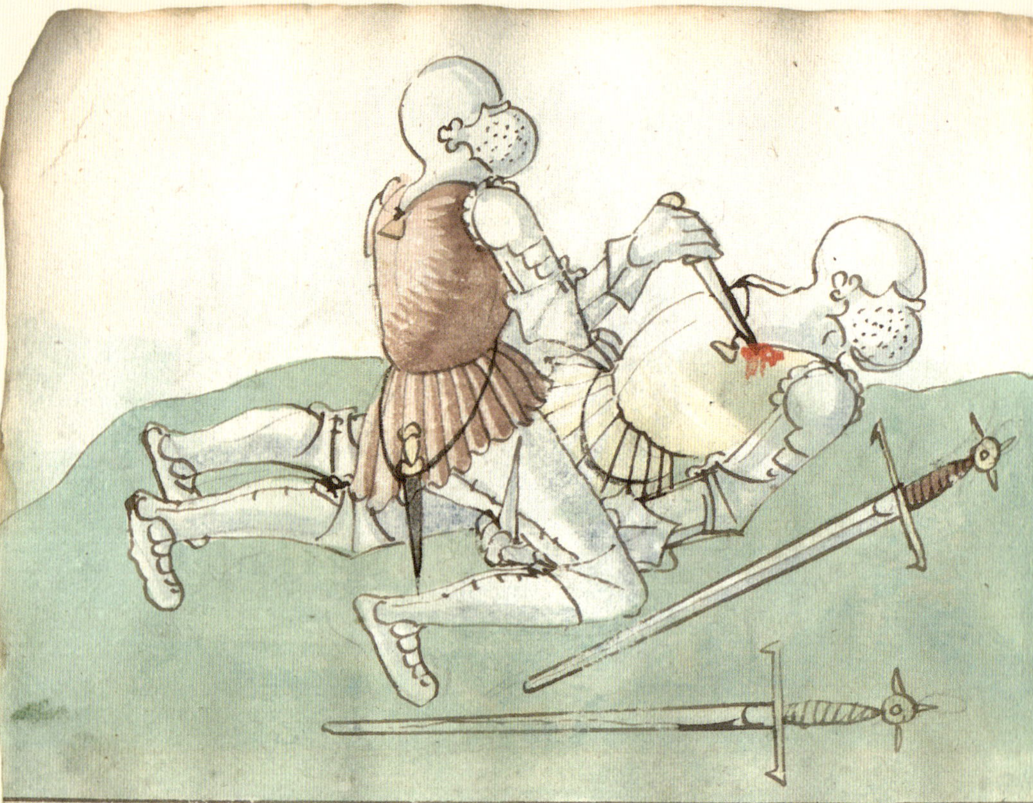

Das zwayundvierczigist Stuck

Item So du empfindest das dich ainer an den rucken
geworffen hatt so stoss mit deiner rechten hand oder mit
deiner lincken aint welcher es dir am ersten kumen
mag und stoss in an sein rechten elenpogen oder begreiff
in halt in vast woll wind dich behend umb so
wirfftu in auff den pauch So sitz behend auff in
und schau wo du in gewinnen magst wie oben
gemalt statt das ist das best stuck im kampff zu
rettung seins lebens

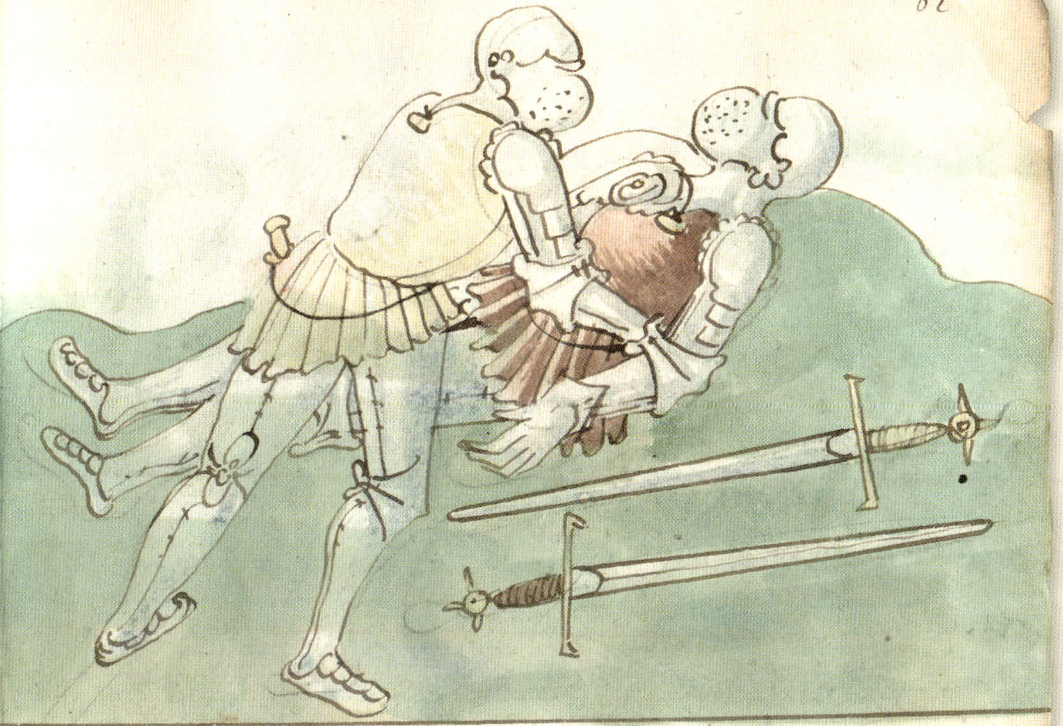

Das drew und viertzigest Stuck

Item So du also auff im sitzst und er will sich under dir auff winden wie vor geschriben und gemalt statt So greiff mit deiner rechten in sein tencken ellenpogen und mit deiner tencken greiff im auff sein rechte So er an dem rucken ligt wie oben gemalt statt So verstu im das er under dir nit auff mag sten du must dein hain wol in hutt haben das er dich nit begreiff er wurd sunst ledig vor dir das seind die besten stuck im kampff

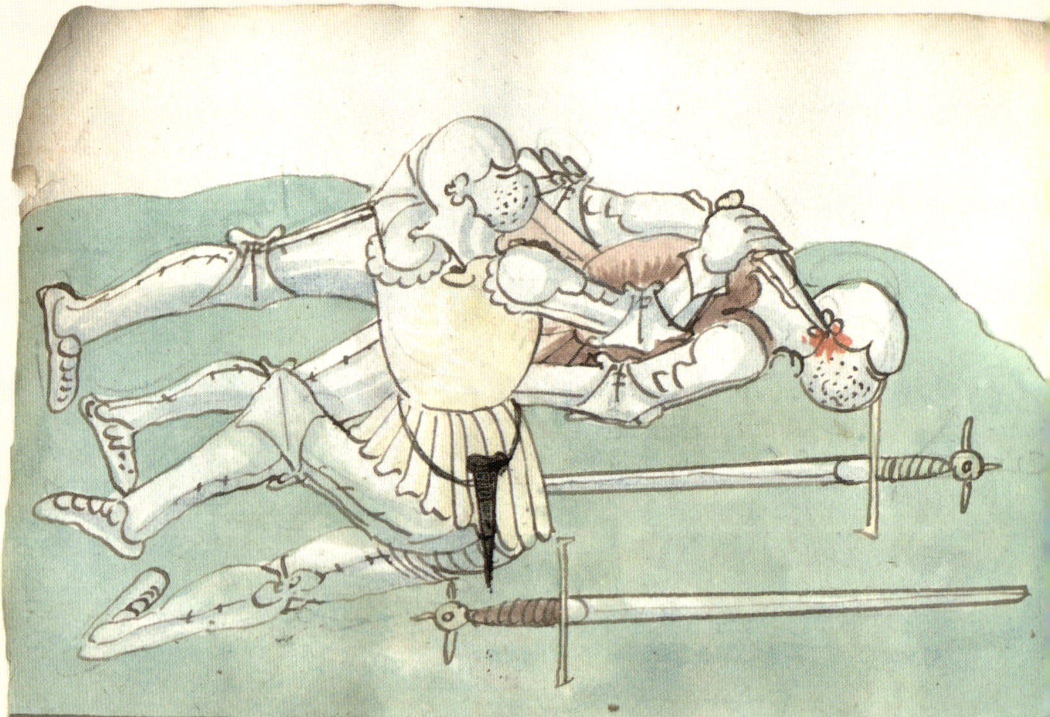

Das vierundfiertzigest Stuck

Item So er sich alsz under dir dir hatt umbgewundn
So heb in vast mit deinen armen auff dein arm
wirff in umb auff den pauch knie hinden mit dein
knie auff sein fersz und stich mit deinem dolch die
plesz oder hastu [illegible] wie du im am nechsten gewin
nen magst wie aber gwalt statt das ist vast ain
gutte arwait

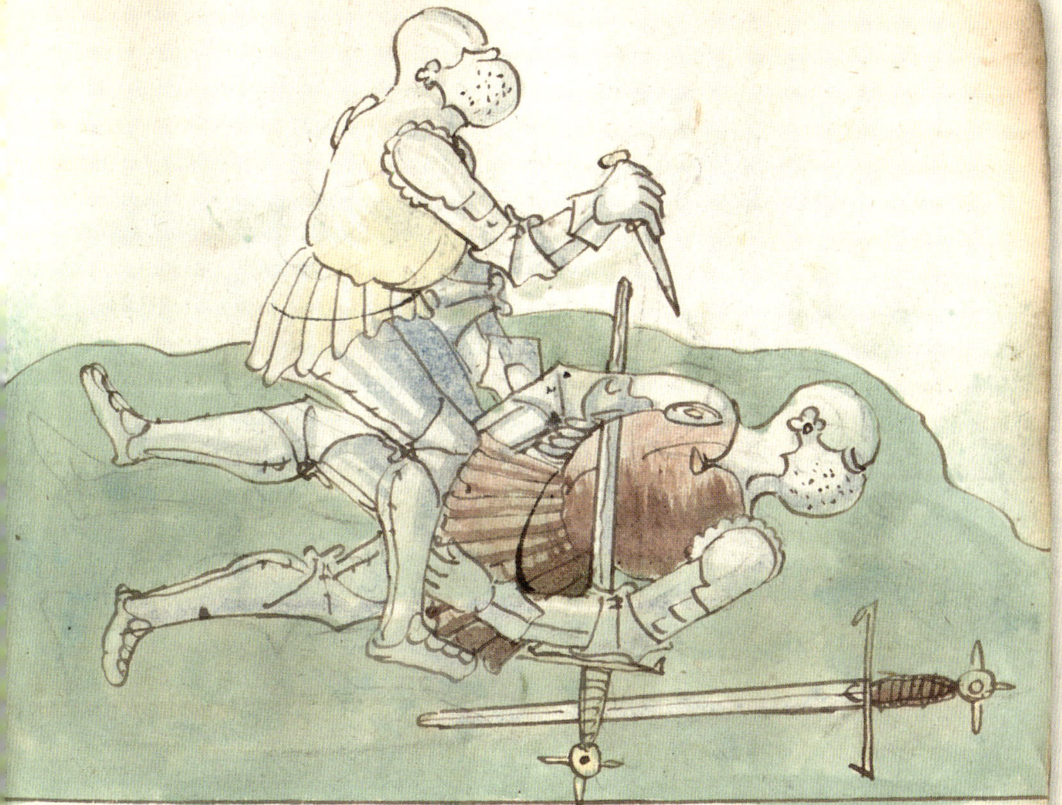

Das ainunddreissigist Stuck

Item So du ainen also umb hast geworffen und kniest auff im und wilt im das leben nit nemen So greiff nach deim oder seinem schwertt welches dir am nechsten ist Stoss ims uber sein rucken under den armen das baid ortt auff dem schwertt ligent [?] So mag er nit auff Sten So magst du dein dolch zucken wie oben gmalt Stett und magst im zu sprechen nach deim gwissen [?]

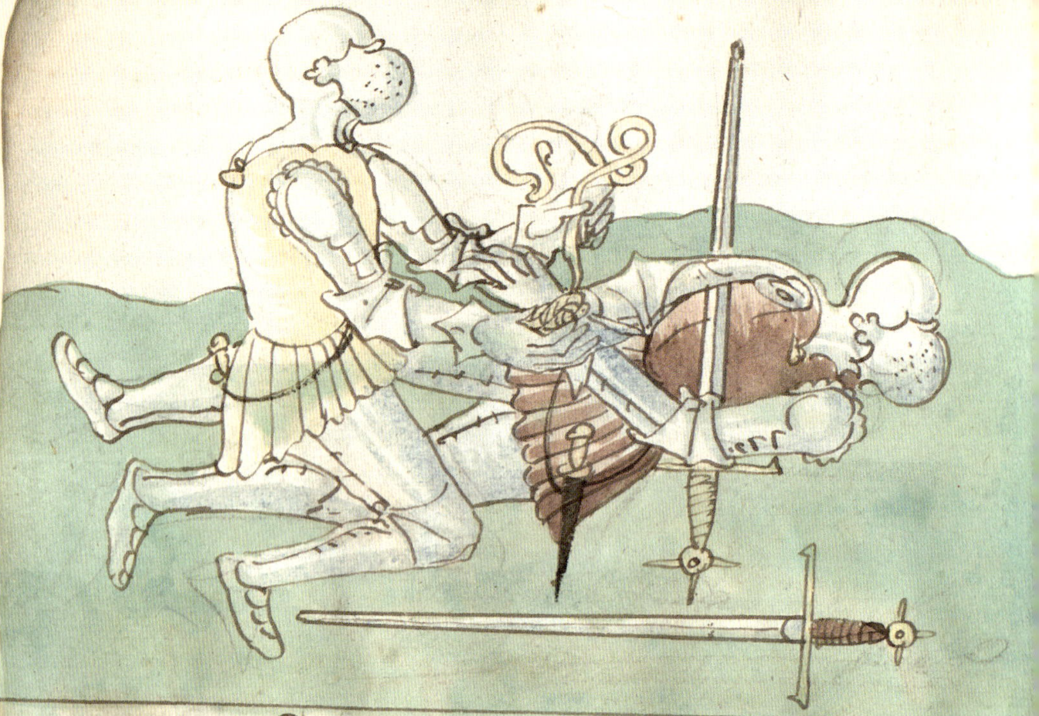

Das Sechsundvierzigist Stuck

Item So du griff wilt tun in der kunst vnd in dem ~~kampff~~ den dich dan der maister gelernt hatt vnd im das schwert durchgestossen hast So schau das du ain kerren oder sterck du oder dein maister bei euch habt vnd bind im die arm zu samen wie oben gemalt stat ~~Dat~~ So mag er nit aufsten du magst auff im kun oder sitzen

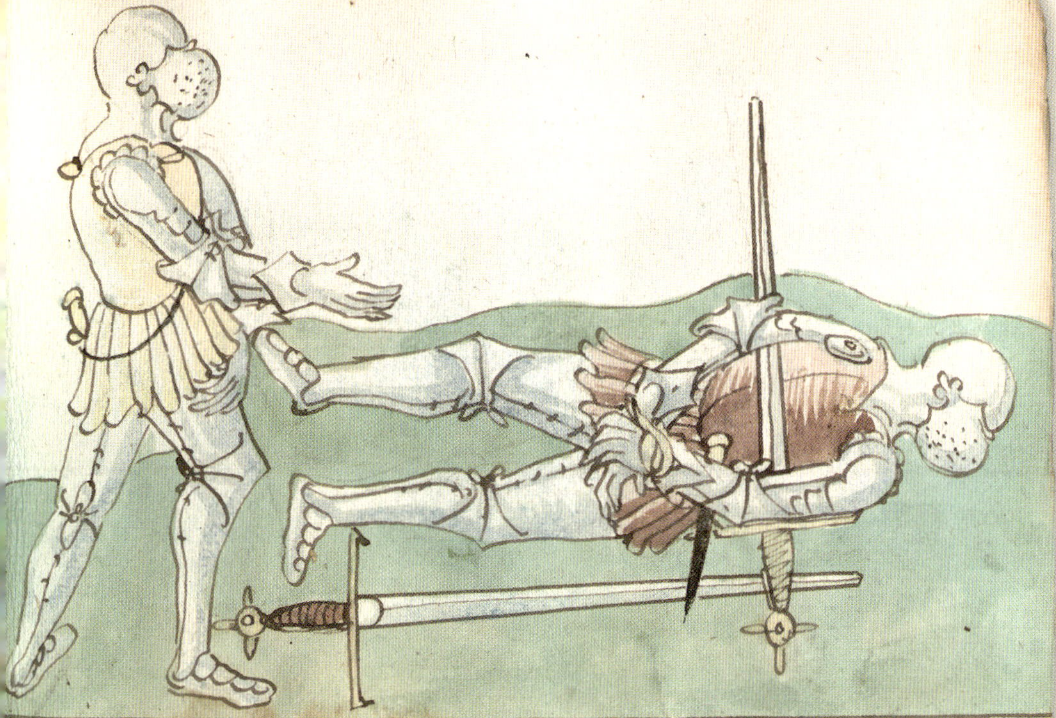

Das Sibenundviertzigest Stuck

Item So du yetz das stuck geben hast und wol gerungen ist und in gepunden hast und im das leben nit nemen wild So lass in ligen gee von im gib im den segen wenn er mag nit auffsten wenn er müed wirtt So wirt er sich wol geben das ist ain beschluss des kampffs

Item vermerk das lest stuck So du klain und schwach bist So schaw so ain starcker mit dir kempft und wil in dich lauffen und sich seiner sterk weist und dich uberstossen wil So setz dich wol in die wag und mach dich klain und ker die seitten für auff welche er laufft So drit du auch auff die ander Seyten

Item So du vermercken wilt vnd wilt die Riterlichen kunst
lernen zu Ross kempffen vnd fechten So sol der maister dich
leren wie du dich halten solst zu veld oder zu holtz kumst du
in der holen weg vnd fürchst du dich So nim dein schwert in die
rechten hand vnd halt in vndersich neben deinem Sattel des
gleichen thu in ainem holtz vnd vor dem kissen fiechst So magst
du erwaiten in dem grüst oder sunst wie dir die arwait
bezaichnet als hernach gmalt stet vnd gar gründtlich ausweist

Item vermerck das erst arwaitten so ainer mit dem spiess auff
dich rent So sich dich wol für ker in die seit vnd was
du in der hand hast es sei ain kolben oder schwert So sleh
im den spiess ab wie du dan hernach sichst das ist die
erst hut vnd ausweisen wie oben gmalt stett

Item So ainer mit dem spiess auff dich rendt und was
du in deiner hand hast schwertt oder kolben so fleuchs
auff und schreit mit deiner rechten hand an den
spiess und nochend bei die sprück in auff dein rechten
seiten das haist die absetzen und ist der ander anfang
in der natturlichen arwait als oben gmalt statt

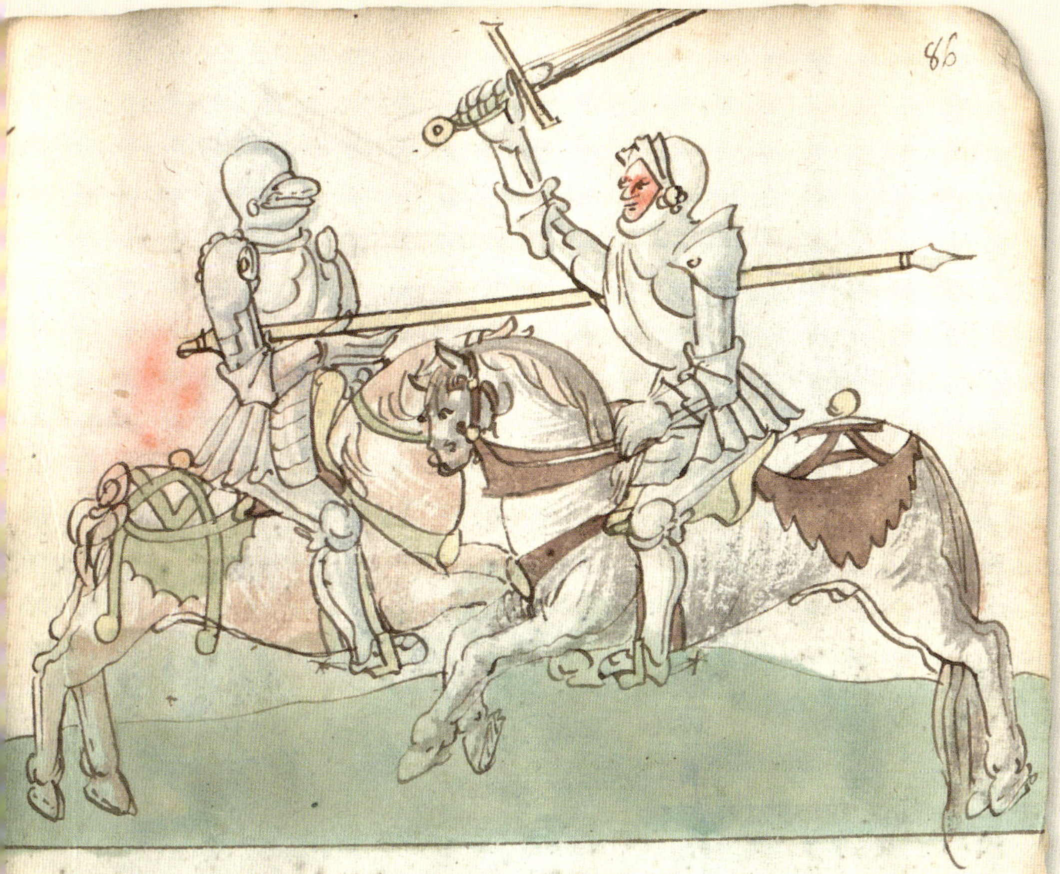

Item So dü dü abgesetzt hast wie vorstett So tritt wol
an in vnd stoss im in das gsicht oder wirff in von dem
gaul wie dü hernach gemalt vnd gschriben vinst vnd bis
vor vürsichtig das dü wissest zü arwitten

Item do zwen zu samen rennent und habent baid spiess so
Schau das du den spiess gar gewiss habest und im rennen lass
den deinen ross auff den kopff und streck das eisen am ren
is ober der ord und so die eissen gleich zu samen kumen
So wechsel unden durch und druck auff dein rechten sitt
So wirfst du in unden an den part oder wa du wilt
Das haist der durchwechsel als oben gemalt Statt

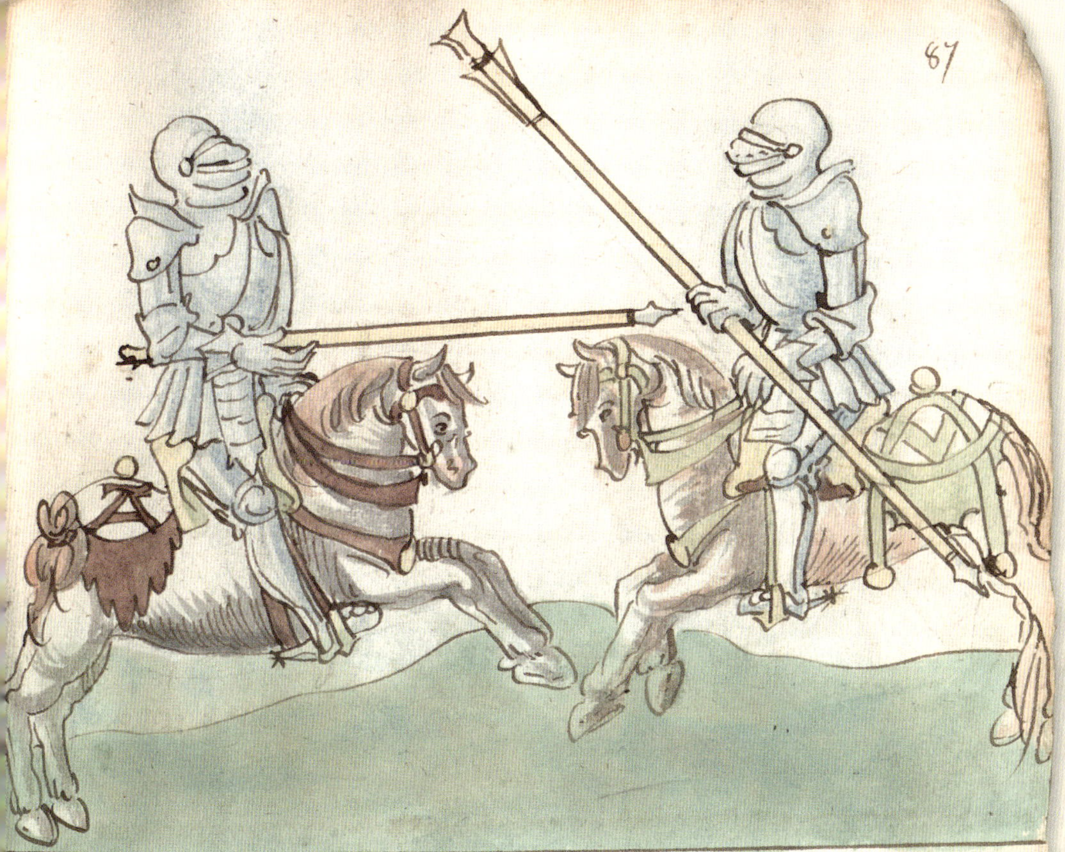

Item So ainer auff dich rent und kanst dein spiess in das
Rist nit bringen machst das du ubereilt wurst so wurff
dein spiess bald auff dein rechten Dyhsen und das eisen
oben der red und setz im den stich ab wie hernach ge
malt stat und bis wol in der hutt

Item So dů den spieß begriffen hast vnd abgesetzt So magst
dů mit dem paustz oder mit dem rysen das haist
nach der kunst auff den ort ·gewart So empfindst
was dů arwaiten solst auch sichst wie aber gemalt
stet

Item so du mit dem pausch abgesetzt hast So würff den
zaum an den hacken den du vor dir solt haben an deinem
gürttel und schau das du das Ross gewaltig fürst mit dem
leib und ge auch mit dem spiess über deinen kopff das
dein rechte hand vor stee und die linck hinden stoss in
mit dem pausch unden an den goder So stost du in von
dem gaul wie vor stett

Item So dir ainer zú eilt und wil dich mit dem paúsch
von dem gaul stossen wie vor gemalt stett so wirff dein
spiess von dir und greiff mit deiner rechten hand in
dein spiess von dir und greiff mit deiner rechten
hand in sein spiess bey dem paústh ruck wol von dir
wie oben gemalt stett So wirffst dú in als bald als er
dich

Item So du sichst das dir ainer in den zausch greift
und hatt dir dein prüch genomen So las nit dein spiess
vall wol in In und greiff mit deiner rechten hand
unden an sein felen trück wol von dir So wirfst du in
von dem gaul mit aller gwalt Stat das haist der
hals Sumerblick

Item So einer auff dich rendt und wil dir dein gaul stechen
So sich dein spiess wol vir dich das er dir nit gerür
und gerür und zeuch auff ain seitten auss mit deinem
gaul so hast im den stich oder stoss abgesetzt wie oben
gemalt statt

Item So du abgesetzt hast wie vorstatt So zeuch wol auff
die seitten halt dein spiess vest So vellt er gehlings
darüber So würffst du in von dem gaul über ain hauffen
wie oben gemalt statt also hat der spiess mit dainer armut
ain end wie wol mer arbait darin ist Sy ist aber nit
als verstendlich

Item vermerck dar oben wie dich das schwertt anfacht
im kampff zů fechten zů Ross die ersten zwai hutten
zů Ross nimpt er die schosshütt wie oben gemalt statt
So schaů das dů dein schwertt aůff die tenngk hüfften
werffest vber dein linke hand das das schwertt ~~ab~~
ab der hand gey das haist die nebenhutt

Item vermerck die ander zwů hůtt So dů sichst das er die
vnderhat hůtt auff seiner rechten seitten vnd senck dein
schwertt wol nider an sein rechten pain So dů sichst die
vnderhůtt So var auff mit deiner rechten handt vber deinen
kopff senck dein ortt wol vnsich gen im thů als wellest
dů im in das gsicht stossen wie sy baid oben gmalt stend
das haist die vnderhůtt vnd die oberhůtt

Item So dir ainer auß der ober hůt zů dem gesicht stost
So ker aůff mit deiner neben hůtt an sein schwertt
So versetzů im den stoß wie oben gemelt statt
daraůß gat vil arwaitt

Item So du mit dem schwertt aufgangen bist und er zuckt und ruckt sein ortt hindersich und wirfft den zornhaw auff dich So fert wol auff mit dem schwertt und fach den straich in das kreuz wie oben gmalt stat darauss gett vil arwait

Item So du sichst das ainer den zornhaw auff dich schlag
en will oder mit dem knopff stossen So far auff mit
deiner vberhutt und setz im an an drei ortten
welches du wilt das merck In dem ainreitten
das erst ainsetzen und dem rechten yechssen das ander
In dem rechten elenpogen in das gelenck des harnasch
das dritt in das gsicht des helms wie oben gemalt
statt und ist der pest gewin

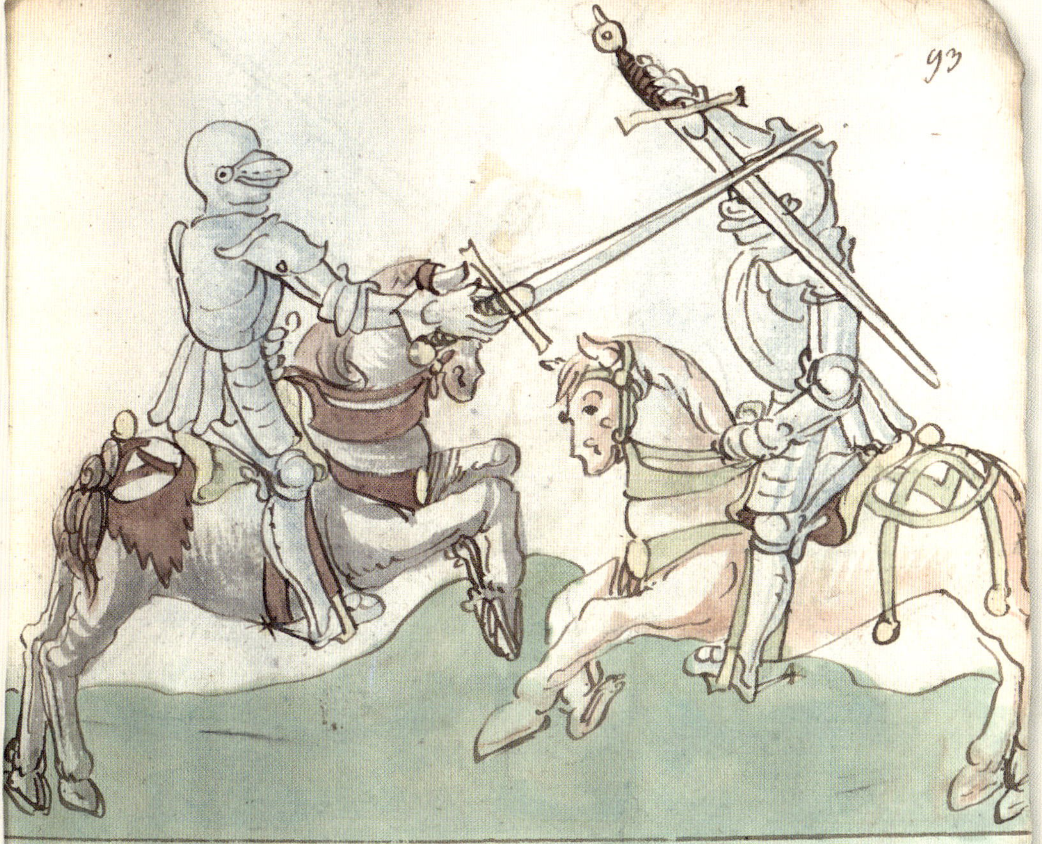

Item vermerck So ainer auff dich rendt vnd will das
ansetzen nemen oder das schiessen So wind wol auff
mit deiner rechten hand vnder dein schwert treib
wol in die höch wie oben gmalt stett So magstu
erwarten was dü wilt

Item So dů das schwertt in der höch gewunden hast so
sol mit dem knopf des schwertt vber dein rechte
hand zeůch wol an dich druck dich wol auff dem
gaůl So prichst dů im den arm oder nimbst im
das schwertt man kan das stůck so gruntlich nit
mallen oder schreiben vnd merck wie oben gmalt
stat So magstů wol in der arwait kůmen

Item So dich ainer also hat vberfallen wöllen So wind dein kopff in die höch vnd das ortt gegen der erden vnd wind wol vber sein arm das schwertt lass hangen in dem mittelhaw vber den kopff legs im an sein halss vnd greiff mit deiner tencken hand in dem schwertt knopf wol nider zeüch an dich So wirffst dü in vor dem gaül deinem zaum wirff an den hacken wie oben gmalt Statt das stuck haist beschliessung bei dem hals

Item So dü Siechst das dich ainer beschliessen weil bey dem hals
So greiff mit deiner tencken hand an Sein Rechte Stoss
wol oberseth wend dich umb ker in den Rucken als
wölest fliechen und schlach mit deinem schwert nach
seinem tenckn pain wie oben gmalt Statt sam das
haist der kastherhaü

Item So du sichst das sich ainer umbkert in die flucht
So lass in nit umbkern fil im In stoss in aus der obern hut
nach dem gsicht greiff mit deiner lenckhen hand auff sein
rechte kurtz in sein hand und schwertt nider als da gemalt
statt werff den zaum an den hafften sie dich den du haben
solt vor dir

Item So ainer In dich Reit vnd wil dir nach dem kopff
Schlachen So dreffstu mit ainem zucken var mit deiner
Rechten hand in sein Rechten Elenpogen Reit wol In In
truck in wie sich wol von dir wer da gwalt Start
So wirffstu in von dem gaul in besser sicherhait so
Schlach mit deinem fuss vnden an dein Rechte pain
vnden an den sporn heb wol vbersich So hilfft du Im
zu dem wurff

Item So du sichst das dich ainer also werffen wil oder mit
dem knopff stossen oder auschen wil was arwait er auff
dich treibt So Rit wol In In und greiff mit deiner
Rechten hand zwischen sein tencke hand durch den hals
vor Im In das gsicht zeuch in In die höh bey seiner
schollen das haist der Sunnenblick als oben gmalt
staet

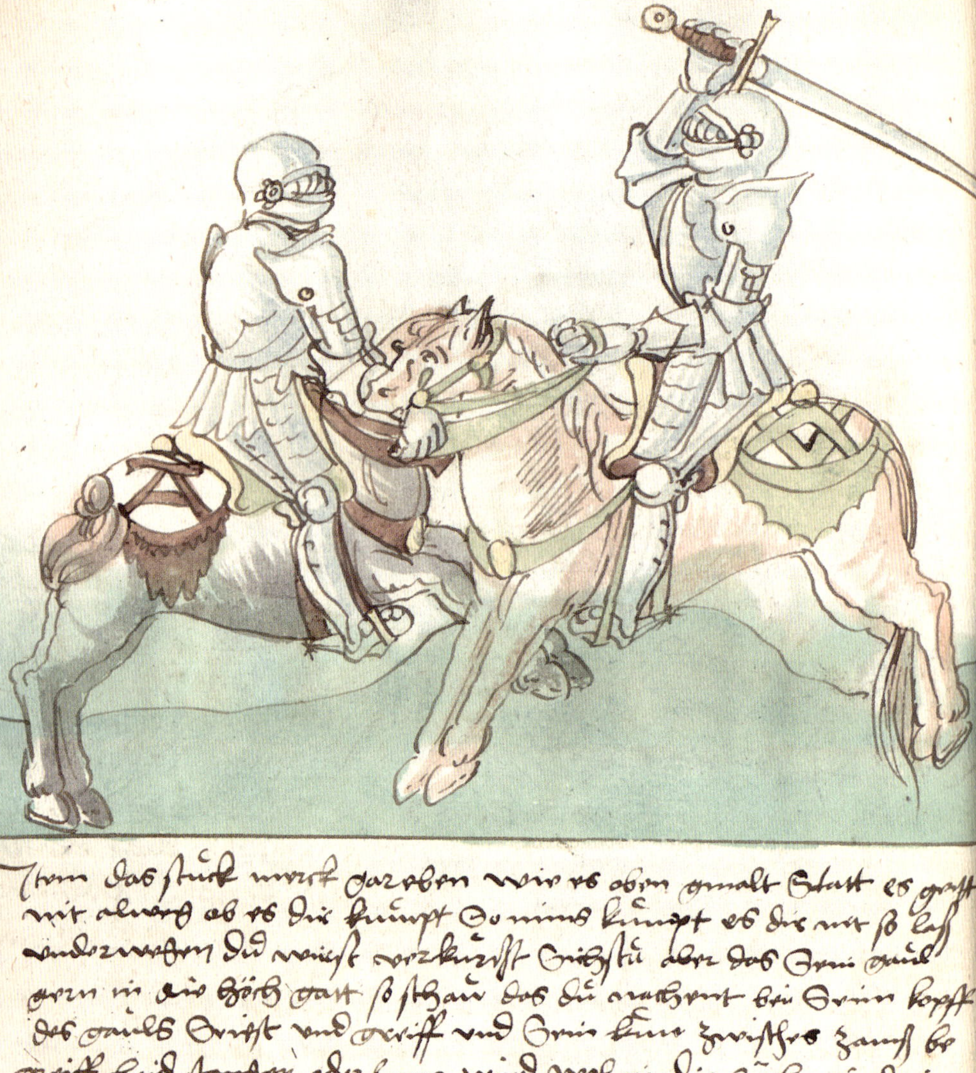

Item das stuck merck gar eben wie es oben gemalt Statt es gatt
nit allweg ob es dir kumpt So nim kumpt es dir nit so lass
underwegen du weist verkürst Sichstu aber das dein gaul
gern in die höch gatt so schaw das du anthert bei deim kopff
des gauls greifst und greiff und dein linck zweischte hand be
greiff baid stangen oder zam wind wol in die höch mit deiner
rechte hand So wirffstu den gaul und den man über ain
hauffen du must gar eben auffmercken ob du es bekumen
magst oder nit wan du es begreiffst So zeuch in auff ain
Seiten wol zu sich

Item So ainer in dich rent und greift dir nach dem zam So
greiff mit deinem rechten arm auff sein tencke achssel und
schau das du im bey dem hals und haubt under wol farest
und ruck wol zu dir und mit dem rechten fuss kum under
an sein rechten fuss mit dem sporn wol hindersich und truck
oben wol dir So wirfstu in von dem gaul oder legst in
vor dich auff den Sattelbogen wie oben gemalt Stat das
Stuck haist die pressen

Item vermerck den pruch auff das vorderstuck So er dir
pressen auff dich genomen hat So var mit deinem rechten
arm vnd dich vber sein rechten vnd greiff mit deiner
lencke hand vber sein lencke das du baid zam zu samen
macht bringen in die hand magst du es nit bekomen
So halt die lencken hand vest vnd reit vnd sich ain
wenig auff die seitten so wiefst du in vor dem gaul
wie oben gemalt statt ist vast ain gutter pruch

Item So er in die flucht kumpt vnd ainer greift dir hinden
in dein goller oder auff dein achssel haist auch dir prissen er
mit seiner lincken hand vnden durch sein rechte arm vnd
wind oben über sein rechte mit deiner lincken hand bück dich
wol nider vnd halt sein harn in deiner rechte hand so prichst
du im den arm oder wirfft in von dem gaul wie oben gemalt
stett es hatt gar gute arwait du must schawen das du wol
nicht sayest in der kunst

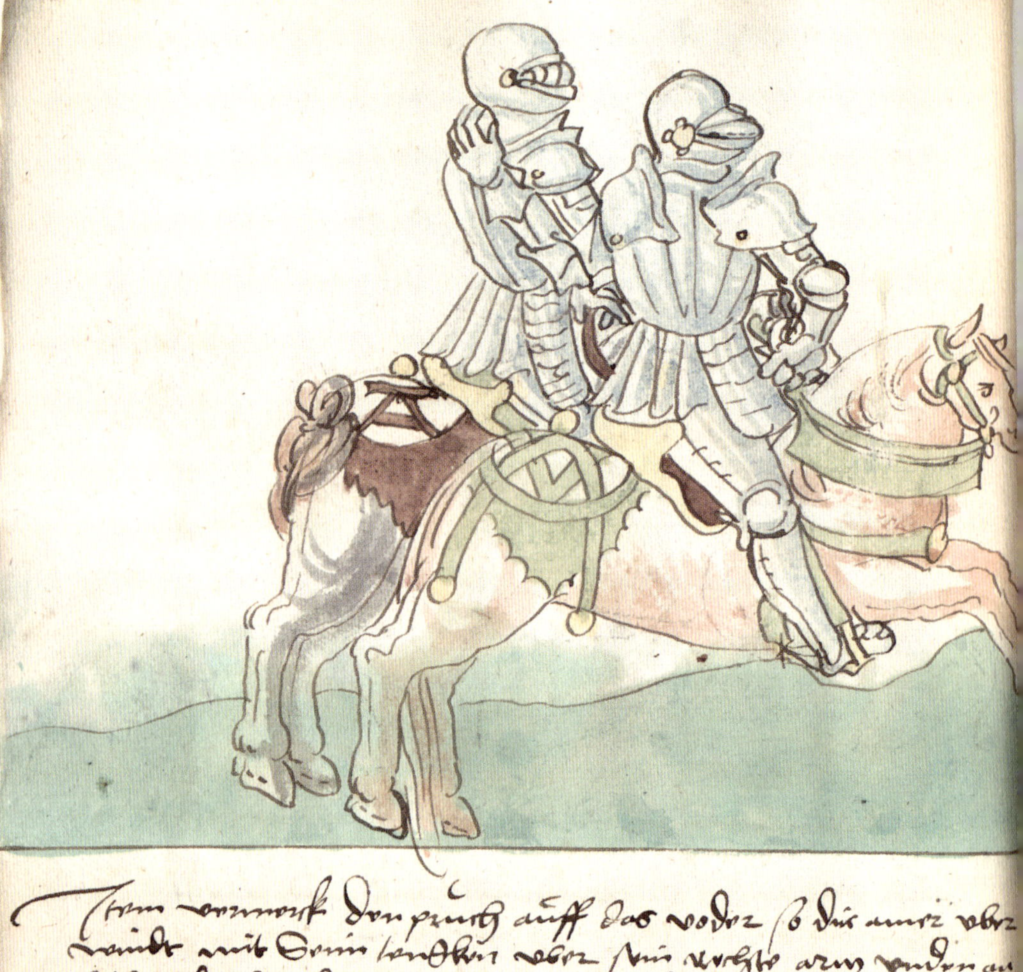

Item vermerck den pruch auff das vorder so dir ainer vber
raindt mit seim tenckhen vber sein rechte arm vnder an
dich vnd fal bald oben vber sein tenckhe nachend bey den vech
ledert dich weg wie sich zeuch auss auff die seiten das
kan man nit machen vnd merck wie der pruch oben
gemalt stett So wirfstu in hintersich auff den gaull

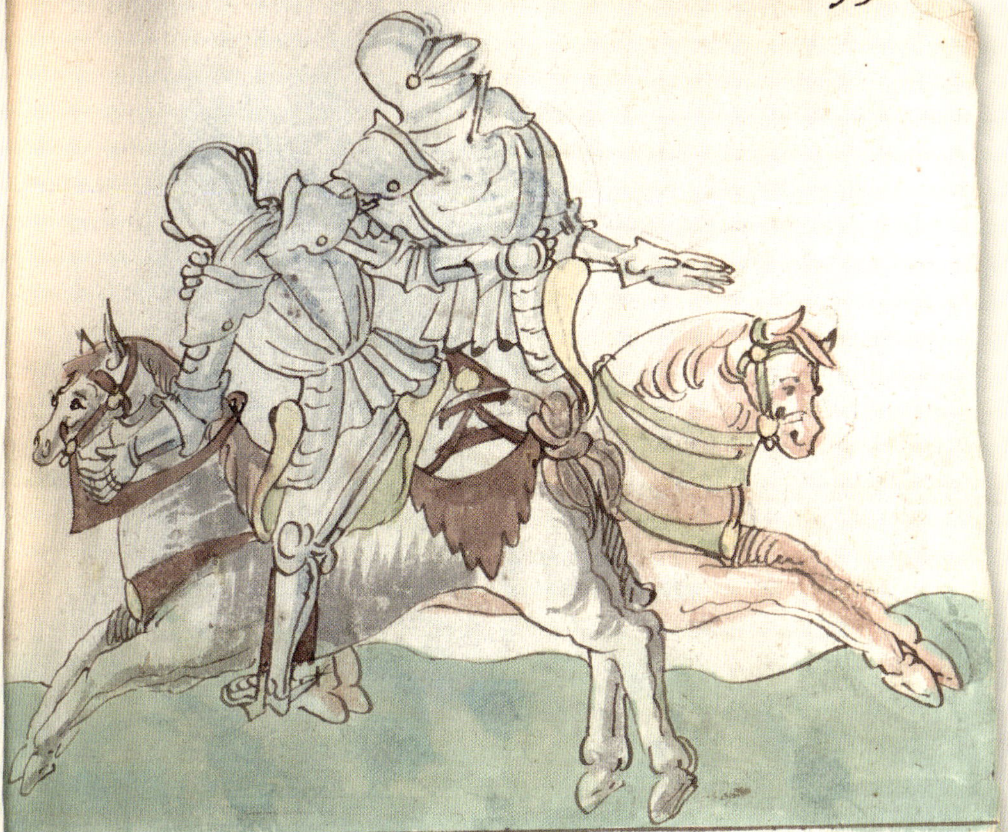

Item wer ain außer bruch und stuck So ainer in dich rendt
und nympt dir persson mit seiner rechten hand auff dem rechten
achssel und will dir halb zwischen nemen und will dich ober
sich treiben in den wurff hindersich So fall bald mitt
dem rechten arm oben uber sein rechte und begreiff sein
rechte bey dem zaum zeuch wol an dich wie du gewalt
hast So wirfstu In

Item vermerck den pruch oben darauff dan es statt gar
zierlich nach ein ander So er dir den zaum begriffen hatt
mit seiner rechten So sal mit deiner tencken auff sein rechte
hand so er hindersich ziehen wil So halt in vest vnd
wind mit dein rechten arm vber sein baid arm in den
schopff oder man deines gauls wie oben gemalt stat greiff
wol auff ain seiten vor im So wirfst in vor dem
gaul

Item vermerck das stuck das haist das prust stuckh und hatt drei pruch
So er dir nach der prust greift und wil dich auß dem sattel heben
So sol mit deinem rechten arm über sein rechte in den zaum zu dem
lincken hand So hastu im den pruch gethon

Item vermerck den andren pruch so du empfindst das er sich begert
und wil dich von dem gaul werffen in der zeit so ledig dein steg
raiff und halt dich vest an im und sitz vor deinem gaul hinder
in wie oben gemalt statt So fürst du in wo du hin willt du
müst in mit baiden henden umfachen in seinen zaum greiff
das haist die schrenckung

Item so ainer auff dich rilt vnd posrett dein so er auff
dich schlecht vnd du magst zu kainer wer nit kumen
So ve auff mit deiner rechten hand an sein schwertt greiff
vnsich in sein schallern oder eisenhutt truck wol hind
ersich vnd tritt wol vnsich als da gemalt statt das
Stuck haist der Sunnenblick So wirfstu in von dem
gaul oder prichst im den hals

Item So einer auff dich eilt und kanst zu wer kumen so schau
das du das anwerffen wol bey dir habst wie dich der maister
lernt wan die stuck sellen verborgen sein man schreibt
sy nit deutlich das anwerffen wo du in triffst reitt
nun von im So wirffst du in von dem gaul wie oben
gemalt statt

Item vermerckh das stückh das hatt ain vesten es hatt
der mit und hat zu erst wenn es die konnen mag stuck
wa dü wellest in der arbait in terent und schlachen ist
es vast güt und im streitten und anderst schain
das dü ehrend begreiffest oben in dem ~~sch~~ schopffgriff
ob las in ni ain halss haucken So ist er ain halber
man So machstü erwarten was dü wilt dü vinderst
vor und nach und haist der schopffgriff wie oben
gmalet ist

Item Jörg Wilhalms Hutters kunst zu auffspring

Wen aine also auff dich reit mit dem spiess wie oben gemalt statt
So schick reck ober sich in die höch so setzt im sein stich ab und
wie die hernach geschriben wirdt

So du im also abgesetzt hast lug das du greifst mit der ar
wart spiest wind oben über sein spies auff dem schwert
oder was du wilst so nimst im sein spies oder ein schwert
des selben glacht oder was jeder am vor wort

Jörg Wilhalm

LONG SWORD

1R [1]Jungk Ritter lerne gott lieb haben vnd frawen | Eren Red frawen wol vnd bis manlich da | man dan sol hiett dich auch vor liegen | dartzu las dich nit betryegen trag aines | manes mutt gen dem der dir vnrecht | thutt vnd setz dein sin in ein herliche sach | vnd lern dartzu manliche Riterschaft doch | ~~Schimpffen vnd schertzen~~ mit freuden geben | die stain stossen stangen schieben vechten | vnd Ringen dantzen vnd springen stechen | vnd turnieren darmyt sol man schene frawen | hofieren doch schimpffen vnd schertzen | vechten wil haben das hertze welch*er* | dar schricket gere [!] kain vechten sol er nit | Lernen warum wer verlorn die kunst | vom schwertt gatt der dunst [!] vnd auch die | grose straich machent die verzagte hertzen | waich darum hörtt man gar vil sagen | Das die vertzagten oft werdent gschlagen | vnd ist ein doretter Sin das ein Nar | mitt vechten wil kumen hin vnd kunst | nitt glernet hatt das klag ich mich | Jörg wilhalm huttmach*er*[2] von aug- | spurg frie vnd spatt

1V [Blank page, only carrying a stamp: 'Bibliotheca Regia Monacensis'.]

2R **Wer auff dich wil hawen | versetzen sol s [!] dich frawen | vnd wen du hast versetzt | So send all sein hew geletzt**
Die zwen stond all beid | gerecht in dem oberhaw
Wilttu kunst schawen so bis lingk vnd Recht mit | hawen vnd lingkh mit Rechtem ist das du starckh be | gerest zuvechten wer nach geitt mit hawen bedarff | Sich kunst wenig frawen haw nachent wie du wilt | kain durchwechsell kumpt an deinen schilt zu kopff zu | leib das vechten nit vermeid mit gantzem leib vicht | was du starckh begerest zutreiben hernach was du | schlächst frisch auff

1 Note in pencil from a later hand – quite likely a librarian's: 'cf. Cgm 3712 f. 65'.

2 Curiously, GE, dated 1533, does not have the name 'Jörg wilhalm huttmacher', but rather a completely different one in the very same spot: 'gregory erhartt'.

Young knight, learn to love God and to honour women. Speak well of women and be valiant when necessary. Beware of lying and don't allow yourself to be deceived. Show gentlemanly courage against those who do you injustice. Set your mind to glorious things and learn also brave chivalry, giving with pleasure. Putting the stone, pushing staffs, fencing and wrestling, dancing and jumping, jousting and tourneying: With these you shall court beautiful women, also with playing and joking. Fencing requires a valiant heart. He who is easily startled shall not learn to fence because otherwise the art of the sword would be lost, as rumour has it. Furthermore, mighty strikes soften timid hearts, that's why one hears quite often that timid fighters are defeated frequently. It is silly to think that a fool will succeed with fencing when he hasn't learnt the art. This is what I, Jörg Wilhalm, hatter from Augsburg, lament from dawn till dusk.

He who wants to strike on you, you shall delight in displacing.
If you have displaced, all his strikes are foiled.
These two both stand in the correct way in the upper strike.

If you want to observe the art, be left and right with strikes. And left with right means that you want to fence strongly. He who follows with strikes may take little pleasure in the art. Strike closely as you wish, no changing through comes at your shield. Don't avoid fencing to either head or body. Fence with the entire body whatever you want to perform strongly hereafter, whatever you boldly strike.

2V **Die zwen stond baid glingks in dem oberhaw**

Vicht nit oben glingks So du gerechts bist und ob | du glingks bist vnd im Rechten auch Ser hinckst | vor vnd nach die zway ding sind aller kunst Ein vr- | -sprung schwech vnd sterckh indes wartt dein ar | wait damit merckh so magstu lernen mit kunst | enn arwaitten vnd wer erschrickt gern der sol kain | vechten lernen

3R **Der stat im oberhau**

Das ist der zornhaw von | der Rechten achseln

Das ist vom zornhaw wer dir oberhawt so nim | den zornhau starckh auff dein achseln vnd haw in | schlecht hinein vber die versatzung und verheng | vnd wind mit dem schwert als welttestu dich zwe- | -rchen vnd lass abfallen zu baiden seitten mit halben | vellern vnd schlag in mit der langen schneid nach | dem Rechten or zu dem kopff hinweg

3V **Der statt im oberhaw**

Das ist der zornhaw von | der glingken achseln

Das ist der zornhaw auff der lingken Seittenn | vnd den selben nim von der lingken achseln vnd | haw im schnel hinein lang zu dem kopff vnd | verheng und wind damit vnd lass aber die veller | lauffen vnd schlag aber wie vor gschriben stat | vnd hindersich hinweg

4R **Das ist der zornortt**

Der statt im oberhaw

Das ist die ausrichtung vnd merckh wer dir oberhawtt | das du im mit dem zornortt eindringest vnd wirtt | er des gwar so nim es oben ab on gfar vnd bis starckh | herwieder wind haw vnd stich sicht ers so nim es nider | das eben merckh haw stich schwech vnd sterckh vnd | nims mit halben vellern vnd schlag aber hinein | wie vor vnd hindersich hinweg

These two both stand in the left upper strike.

Don't fence above left if you are right. And if you are left you are greatly hampered on the right. Before and after, these two things are the origin of the entire art. Weak and strong, *Indes*, consider your work with this, so you can learn to work skilfully. He who is easily startled shall learn no fencing.

He stands in the upper strike.
This is the *Zornhaw* from the right shoulder.

This is about the *Zornhaw*. When he strikes above to you, then take the *Zornhaw* strongly on your shoulder and strike directly into over the displacement. Hang and wind with the sword as if you wanted to make a *Zwerch*. Let it drop off to both sides with half *Fehlers* and strike with the long edge to his right ear or away to the head.

He stands in the upper strike.
This is the *Zornhaw* from the left shoulder.

This is the *Zornhaw* on the left side. Make this one from the left shoulder. Strike quickly long towards his head, hang and wind at the same time. But let the *Fehler* run off and strike again as is written above and away backwards.

This is the *Zornort*.
He stands in the upper strike.

This is the explanation. Note: When he strikes an upper strike to you so that you enter with the *Zornort* and he notices it, then take it off above without danger. Be strong against, wind, strike and thrust. If he sees it, take it down. Pay close attention to this: Strike, thrust, weak and strong. Take it with half *Fehlers* and strike into as before, and away backwards.

4V **Der stat im oberhaw**

Das ist der ander[3] zornortt

Das ist der ander zornortt auff der andern seitten den | nim auch also wen sy gand baid wie der zornhaw | von der achsel vnd mus einer aus einem haw | ain stich machen vnd muss den haw verwerffen | vnd das nim auch flux hinweg wye das ob*er*

5R **Hie vacht an der krieg**

Die zwen stond in dem | krieg

Das ist die ausrichtung vonn dem krieg wer des | kriegs oben remet der wirt oben nider gschemett indes | vor vnd nach dem krieg Sey dir nitt gach nim | des in allen sachen wiltu den krieg machen | gloss merckh

5V **Das stuckh ist in dem krieg**

Das ist auch von dem krieg vnd solt auch wissen | das in allen winden haw vnd stich schwertter | vberlauffen schwertt auswinden in allem treffen | wiltu den maister äffen wan es sind gar vil | brich darub*er* vnd auch dartzu nim war wie | einer mit dem krieg gebar gloss merckh

6R **Das ist ein auswinden vnd bruch von dem krieg**

Das ist aber ein stuckh von dem krieg vnd ein bruch des | vordern stuckh vnd ist ein auswinden das geit haw / | stich gutt schneiden finden auch soltu mit brauchen | haw stich schneid vnden in allem treffen wilttu | den maist*er* äffen

6V **Das ist auch von dem krieg**

Das ist auch von dem krieg schulch zu dem ortt | vnd nim den hals one forcht auch solttu ver- | -wenden krum vnd sich vmb vnd auff vnd | brieff im sein gfertt ob er sey waich oder hertt | in sein ~~st~~ gfertt gloss merckh

3 The word has been inserted above the line.

He stands in the upper strike.
This is the second *Zornort*.

This is the second *Zornort* on the other side. Make it in the same way, for both are made like the *Zornhaw* from the shoulder. One has to make a thrust out of a strike and has to throw forward the strike. Take this away quickly like the one above.

Here begins the *Krieg*.
These two stand in the *Krieg*.

This is the explanation of the *Krieg*. He who aims for the *Krieg* above will be put to shame above, below. *Indes*, before and after, don't hasten with the *Krieg*. Use it for everything when you want to make the *Krieg*. Note the gloss.

This piece is about the *Krieg*.

This is also about the *Krieg*. You shall furthermore know that in every winding, there is a strike and thrust, running over swords and winding out the sword in every encounter if you want to fool a master, for there are many breaks against it. To this end, pay attention to how one moves with the *Krieg*. Note the gloss.

This is a winding out and a break from the *Krieg*.

This is another piece from the *Krieg* and a break against the previous piece. It is a winding out which gives strikes, thrusts and lets you find good cuts. You shall also employ at the same time strikes, thrusts and cuts below in every encounter if you want to fool a master.

This is also about the *Krieg*.

This is also about the *Krieg*. Squint to the point and take the neck without fear. You shall also use the *Krump*. Look around and up and check in his encounters whether he is soft or hard in his encounters. Note the gloss.

7R **Das ist auch von dem krieg**

Das ist von dem krieg die ausrichtung wer also ainen | vacht bey dem halss mit dem schwertt der mus | nemen mit eim durch winden vnd wider winden | das geit gutt schwech vnd sterckh vnden vnd verwend | dein schwertt an dem halss vnd zuckh den lingken fuss | hindersich vnd schneid in ob dem halss gloss merckh

7V **Das sind örtter in dem krieg**

Das sind die valschen örtter in dem krieg das gatt | ernstlichen zu wen warum es geherend vil durch | geng dartzu vnd winden vnden durch oder wind | oben vber behend vnd wirff den ortt auff die | hend vnd verker zwingen durchlauff auch wiltu | Ringen

8R **Das ist auch von dem krieg**

Das ist der krieg wan du einen schlechst in die Rechten seit*en* | so nimpt der ander den krieg vnd bedarff dir nit mer | dan dein blöss versetzen mit der langen schneid vnd so | scheust er dir mit dem schwertt oben vber vnd thu | als du es gmalt sichst

8V **Das ist auch von dem krieg ein schwertnemen**

Gloss merckh das ist ein schwertnemen vnd get auss | dem krieg das nimpt einer gar gering hinweg | das es sind noch gar vil gutter stuckh die zu dem | krieg gehörentt vnd ist des kriegen mangerlay | vnd darum so solttu merckhen gar vil ding

9R **Hie vachen an die | vier blöss**

Das sind zwu blöss

Hie vacht an die blöss zum kopff vnd auch zu der | Seitten die vier blöss wiss zu Remen so schlechstu | gwis on all gefar on allen zweiuel wie er ge | bar das wirstu wol gewar auch kumpt es noch | hinden hernach wie du soltt die vier blöss | brechen gloss merckh

This is also about the *Krieg*.

This is the explanation of the *Krieg*. Who catches someone in this way by the neck with the sword must make a winding through and a counter-winding. This gives beneficial weakness and strength below. Turn your sword at the neck, jerk the left foot backwards and cut above into his neck. Note the gloss.

These are the points in the *Krieg*.

These are the deceitful points in the *Krieg* that are applied seriously because many ways of going through belong to them. Wind through below or wind over above, throw the point on the hands and turn. Enforce running through if you want to wrestle.

This is also about the *Krieg*.

This is the *Krieg*: When you strike someone into the right side, the other one makes the *Krieg*. He doesn't need to do more than to displace your opening with the long edge, and in this way, he shoots over you with his sword. Make it as it is painted above.

This is also about the *Krieg*: A taking of the sword.

Note the gloss. This is a taking of the sword which derives from the *Krieg* and which can be performed quite easily. There are many other useful pieces that belong to the *Krieg* because the *Krieg* has many sides. That is why you shall pay attention to many things.

Here begin the four openings.

These are two openings.

Here begins the opening of the head and also of the side. You shall know how to aim for the four openings, so you can hit securely without danger and regardless how he moves. You'll notice this clearly. Further on it is also written how you shall break the four openings. Note the gloss.

9V **Das sind auch zwu blöss**

Das sind die andern zwu blöss zu der andern seitt*en* | darum so lug vnd merckh auff wie die gand | vnd wer die blöss vnden nemet der wirtt | oben bey den orn gschemett gloss merckh | Schwech vnd sterchk etc

10R **Das ist der krumphaw auff | der Recht*en* Seyten wechsell**

Der statt im ober- | haw

Das ist von dem krumphaw die ausrichtung krum | auff behend wirff den ortt auff die hend haw | krum zu der flöche wiltu den maist*er* schwech*en*[4] | krum nit kurtzhaw dein durchwechsel damitt | Schaw oder verwend wilttu Ringen vnd drit | im in den driangel gloss merckh

10V **Der stat im oberhaw**

Das ist der krumphaw | von der lingken Seitten

Das ist krump auff behennd wer wol versetzt mit | Schreitten er vil hew letzt haw krum zu der vlech | eim wilttu den maist*er* schwechen wan es glitzt oben so | Stand ab das wil ich loben krum schlecht vnd nyt | kurtzhaw den durchwechsel damit schaw haw krum | wer dich Irret der edel krieg in verwirett

11R **Das ist wie man die vyer blöss | brechen sol der ein der schlecht | in mit der blöss in die seitten | So bricht der ander Sy Im | vnd schlöcht in hinden auff | Sein kopff**

Der schlöcht die blöss | dem andern in sein | Rechten seitten

Das ist blöss brechen vnd wilttu dich Rechen die vier | blöss kunstlichen brechen du oben duplier vnden | Recht muttier ich sags die virwar sich bschitzet kain | man one far hastu vernomen zu schlagen mag ir | kainer kumen das hab ich vernomen gloss | Merckh

4 JWA1, fol. 40r: 'äffen'.

These are also two openings.

These are the other two openings on the other side. Therefore watch out and pay attention how they are employed. Whoever takes the openings below will be put to shame at the ears above. Note the gloss. Weakness and strength.

This is the *Krumphaw* on the right side. Changer.

He stands in the upper strike.

This is the explanation of the *Krumphaw*. Make the *Krump* upwards nimbly, throw the point on the hands. Strike *Krump* to the flat if you want to weaken a master. Don't strike the *Krump* shortly, look out to change through with this. Or else turn it if you want to wrestle. Step him into the triangle. Note the gloss.

He stands in the upper strike.

This is the *Krumphaw* from the left side.

This means: make the *Krump* nimbly upwards. Who displaces well with steps impairs many strikes. Strike *Krump* to the flat if you want to weaken a master. When it clashes above, then stand back which I will praise. *Krump* plainly and don't strike short. Look out to change through with this. Strike *Krump* against he who confounds you; the noble *Krieg* will confuse him.

This is about how you shall break the four openings. One strikes the other to the opening in the side. Then the other one breaks this and strikes him behind on his head.

He strikes to the opening in the other one's right side.

This is about breaking the opening. If you want to retaliate and break the four openings skilfully, then use the *Duplieren* above and the *Mutieren* properly below. I tell you indeed that no one protects himself without danger. If you have understood this, he will be unable to strike. This I have understood. Note the gloss.

11V **Da bricht diser die blöss | auff der andern seitten | vnd ist auch die blöss | brechen**

Der schlöcht die blöss | dem andern in sein | lingken Seitten

Das ist die blöss brechen auff der andern seitten | vnd brich sy wie du Si gmaltt sichst du must | auch oben dupplieren vnd vnden Recht muttier | renn Item die blöss brechen soltt gmalt sten | daruorn vor dem krumphaw

12R **Die hutt vom tag**

Der Alber[5]

Item das sind zwu hutten vnd ist das die hutt | vom tag der sein schwertt vbersich hebt vnd | der Ander der sein schwertt vndersich hebt bei dem | lingken fus das ist der Alber vnd haist auch | die Eisin portt dan es sind vier hutten ochs | vnd pflug das sind auch zwu hut-ten gloss merckh

12V **Schaittelhaw**

Ein leger oder die Eisinn | portt

Item das ist von der kron etc vnd der der das schwertt | vbersich hebt der statt in dem schaitelhaw vnd der | ander in eim leger vnd wan der schlecht vn dachab | so nimpt der ander die kron also schon gloss | merckh

13R **Das ist der zwerchhaw**

Der statt im ober | haw

Das ist die[6] zwerch benimpt was von tag herkompt die | zwerch mit der sterckh dein arwait damit merckh die | zwerch zu pflug zu dem ochsen gehertt fug vnd ~~w~~ | was sich wol zwercht mit springen dem haupt ge | fertt vnd was veler veruarn von vnden auff | zwiuach[7] nach wunsch Sy Rerentt gloss merckh

5 Note in pencil: 'Frank'.

6 The letter 'd' superscribes a previously written 'z'.

7 'iu' is actually 'uu' with a tittle.

Here he breaks the opening on the other side, which is also about breaking the openings.

He strikes to the opening in the other one's left side.

This is about breaking the opening on the other side. Break it as you can see painted above. You must use the *Duplieren* above and the *Mutieren* properly below. Item: Breaking the opening will be seen later on before the *Krumphaw*.

The guard *vom Tag*.

The *Alber*.

These are two guards. One is the guard *vom Tag*, shown by the one raising his sword up, and the other, shown by the one holding his sword down next to the left foot, is the *Alber* which is also called the *Eiserne Pforte*. For there are four guards: *Ochs* and *Pflug* are also two guards. Note the gloss.

***Scheitelhaw*.**

A position or the *Eiserne Pforte*.

Item: This is about the *Kron*. He who raises his sword upwards stands in the *Scheitelhaw*, and the other one in a position. When one strikes from above the other one already uses the *Kron* for his advantage. Note the gloss.

This is the *Zwerchhaw*.

He stands in the upper strike.

This means: The *Zwerch* takes away what comes from *vom Tag* (from above). Make the *Zwerch* with strength, and at the same time consider your work. *Zwerch* towards the *Pflug*, rightly to the *Ochs*. Whatever makes the *Zwerch* well with jumps endangers the head. What the *Fehler* brings up from below wounds twice at will. Note the gloss.

13V **Das ist die zwerch auff | der andern seitten**
Der statt im ober | haw
Das ist die zwerch auff der andern seitten den | nim auch wie auff der ander Seitt*en* verker | zwing durchlauff vnd auch Ring vnd in dem | Elenpogen in der wag vnd dritt in den lingk*en* | vnd bis nit lass in baiden zwerchen magstu | Ringen gloss merckh

14R **Der bricht die zwerch**
So schlecht der die | zwerch
Das ist das stuckh wie einer die zwerch | brechen sol da bricht ein zwerch die andern | mit auff sitzen vnd mit vnderschiessen vnd | mitt vberschiessen daraus magstu nemen | die zway hengen ob du wiltt dan es sind | zway hengen die den man seer Engen | gloss merckh

14V **Der stett in der hutt von | tag vnd wertt sich als | vil er mag**
Der spricht also was sich | wol zwercht mit springen | dem haupt gefertt das | ist das stuckh
Das ist auch ein stuckh von der zwerch | was sich wol zwercht mitt springen dem | haubt gefertt auff ein seitten zu der andern | ist auch eins das ist besser den das also | vernim gloss merckh

15R **Der leit in der zwerch vnd hatt die | genomen mit eim sprung vnd | Stost im den arm hinweg von | Im**
Das ist auch von der zwerch das nimpt man | auch mit eim sprung das nim also wie du | es gmaltt sichst darnach kumpt der bruch dar | um sich dich vir dan es ist kain stuckh es | hatt ein bruch darum merckh kunst ver | nim Sy schon

This is the *Zwerch* on the other side.
He stands in the upper strike.
This is the *Zwerch* on the other side. Make it as on the other side. Turn, enforce running through and wrestling; in the elbow, in the scales. Step to the left and don't be lax. You may wrestle in both *Zwerchs*. Note the gloss.

He breaks the *Zwerch*.
This is how he strikes the *Zwerch*.
This is the piece about how one shall break the *Zwerch*. Here one *Zwerch* breaks another with placing upon and with shooting under and over. Out of this you can make the two hangings, if you wish. Because it's two hangings that constrain the opponent greatly. Note the gloss.

He stands in the guard *vom Tag* and defends himself as well as he can.
He says this: What makes the *Zwerch* well with jumps is dangerous for the head. This is the piece.
This is another piece from the *Zwerch*. What makes the *Zwerch* well with jumps is dangerous for the head from one side to the other. There's one that is better than that. So understand and note the gloss.

He is positioned in the *Zwerch* and has made it with a jump. He pushes the other one's arm away from him.
This is also about the *Zwerch* which is made with a jump. Make it as you see it painted. After this follows the break. So be careful, because there is not one piece without a break. Therefore you shall pay attention to the art and understand it well.

15V **Der statt auch in der zwerch | vnd den wil hinweg stossen Der aber bricht Im | vnd zwercht sich oben | vber sein kopff**
Das ist der bruch darwider du magst | wol fluchs ein schnit nemen vnd hinweg | das er dich nit erlangen mag den ein stuckh | bricht das ander vnd merckh dabei was dir | gutt Sey

16R **Das ist auch von der zwerch mit | Springen dem haubt gefertt | wie er gebar mit seiner kar | on als gefar etc**
Das ist auch ein stuckh was sich wol zwercht | mitt springen dem haupt gefertt vnd ein schnidt | Ein stich vnd vbergen vnd bey dem hals vachen | Also verstand es wie du es oben gmalt sichst | gloss merckh

16V **Das ist die zwerch oben ein auch ein stuckh**
Das ist aber von der zwerch von dem vber | gen das du es verstandest das gar vil aus | der zwerch kumpt wen man es sucht wen | Sy ist gutt zu allen sachen schwech vnd | Sterckh machen gloss merckh

17R **Das ist der ochss**
Das ist der pflug
Das ist von dem ochsen vnd pflug die ausricht | ung die ding vnd die zwei ding sind aller kunst | ein vrsprung ochs vnd pflug sind dem krieg harten | fug In allen sachen den altten schnidt mit ze | machen so bistu sicher in allen dingen vnd | mag dir mit miselingenn

17V **Das ist der ochs**
Das ist der pflug
Das ist der ander ochs vnd pflug vnd sind auch | dem krieg hartten fug Item den ochsen vnd | pflug sol man treiben von baiden Seitten | Dan Sy haben vier winden vnd vier örtter | die magstu brauchen wie du wiltt wen es | Sind auch zwu hutten gloss merck

through the *Kron*, you break it neatly and hard. Press the strikes, retreat with cuts. Let us make this bravely in all settings.

This is: squint to the point and take the neck without fear and then strongly into the scales.
This one says: a lot of breaks against this belong here that are described hereafter.
This is about the *Schilcher*. Squint to the point and take the neck without fear. Position yourself in the scales and hold on to yourself. If you want to get away from him then wind out and away from him. Note the gloss.

He strikes over above.
He has the neck without fear.
This is again about the *Schilcher* and it is a break against it as you can see in the painting. Notice that there are even more pieces that break it. Pay attention how he moves. Note the gloss.

This is the *Schaitelhaw.*
This is the *Schilcher.*
This is the explanation of the *Schaitelhaw*. This means *Schilcher* with *Schaitler* [are both a danger to the head. The *Schaitler*] is dangerous to the face, with its turn it is very dangerous to the breast. The *Kron* also saves what comes of it. Note the gloss.

This is the short edge.
Schaitelhaw.
This is also about the *Schaitelhaw*. It is important to know that one shall make the *Schaitelhaw* with three steps and one can make it with one step. The *Kron* takes away what comes from it. With the cut through the *Kron*, you break it hard and efficiently. Press the strikes, pull away with cuts. Note the gloss.

23R **Das ein versetzen**

Das ist ein leger

Das ist von den vier legern vnd auch alwegen | ain leger vnd ein versetzen darauff die vier | leger ~~v~~ allain dauon haltt die sag vnd fleuch | die gemain ochs vnd pflug alber vom tag ist | wol sein fug

23V Das ist auch von den legern vnd ist zew- | -issen das ein leger das ander bricht vnd | das sind zwai leger ainer leitt oben der | ander vnden das merckh also vnd ver | nim das der vnderst dem obern wol durch | mag gen vnd gieng im auff sein kopff | vnd darum durchgen thutt woll so ichs | Sagen sol gloss merckh

24R Das ist auch von den legern das der den | auff dem leger schewst krum auff vnd | scheust in vber sein ~~hend~~ schwertt vnd | ist das vberschiessen auch ein leger das | ander bricht gloss merckh

24V **Das ist ein versetzen**

Das ist auch ein leg*er*

Das ist von den vier versetzen die die | leg*er* serr letzen doch vor versetzen hiet | dich gschicht es dir nott es miett dich | vnd bis Resch vnd krum[11] vor hastu ver | nomen zu kainer kunst mag er nitt | kumen gloss merckh

25R Das ist aber von den vier legern vnd | die versetzen darauff vnd vberschiessen | das sind auch leger das merckh da vnd | enn mit der sterckh vnd lang in der | wag vnd wen er dich woltt aufheben | so magstu ein schnidt nemen ob du wilt | oder magst verhengen was dir am best*en* | viegtt gloss merckh

11 JWA1, fol. 25v: 'kom'.

This is a displacement.
This is one position.
This is about the four positions: There's always a position and its matching displacement. You shall only value the four positions and back away from the common ones. *Ochs*, *Pflug*, *Alber* and *vom Tag* befit him well.

This is also about the positions. It's good to know that one position breaks another. These are two positions: One stands above, the other one below. Notice this: understand that the lower one can easily go through the upper one, especially on his head. That's why going through is rather beneficial, if I may so express myself. Note the gloss.

This is again about the positions. One shoots on the one in the position crookedly by shooting over his sword. The shooting over is also one position that breaks another. Note the gloss.

This is a displacement.
This is another position.
This is about the four displacements that impair the positions severely. But beware of displacing; if it happens, it gives you trouble and distress. Come forward swiftly. If you have understood this, he may not come to any art. Note the gloss.

This is again about the four positions and the displacements against them. The shootings over are also positions. Notice this below with the strength and long in the scales. If he wants to lift you up, you can employ a cut if you wish, or you can hang – whatever suits you best. Note the gloss.

25V Das ist aber von den vier legern vnd von | den versetzen da merckh auff vnd schaw | das Sind auch zwai leger vnd das ist | auch ain leger der vberschossen hawt | vnd haben die leger ein End vnd die | vier versetzen Sy sind all bey ein ander | dan es ist ir vil der sy suchtt gloss | merckh

26R Das ist von dem nachRaisen lere zwiuach | in der were vnd zway daraus nim vnd | dein arwait darauss begind vnd brieff | dein gfertt ob Sy sind waich oder hörtt | das selb lere Indas das wortt schneidett | Sere nachraisen zwiuach so nymstu den | altten schnidt mit macht vnd hab acht | gloss merckh

26V Das ist ein anders nachraissen den | das nachraisen das solttu Nemen auff | baiden Seitten dritt lang wilttu schreitten | wen das nachraisen ist mangerlay | vnd das ist das nachraisen wan einer dich | gschlagen hatt vnd wil von Dir gen oder | abziechen So volg im nach vnd wer | dich gloss merckh

27R Das ist von dem vberlauffen da merckh das das | vberlauffen ist so einer den andern aussen | vberlauft mit dem knopff auswendig mit | macht vnd das ist ein ausser vberlauffen vnd | der ist noch mer vnd darum wan einer | dir wil vberlauffen so heb dein schwertt | Starckh in die höch so kan er dir nit vber | lauffen gloss merckh

27V Das ist auch von dem vberlauffen vnd das | ist ein Inwendigs vberlauffen vnd wer vnd*en* | Remett vnd wen dir ainer vberlauft so wirstu | geschemett vnd wen es glutzt oben so setz es ab[12] das wil ich Loben dein arwait mach waich | oder hertt vnd druckh das zwiuach wie du | wiltt gloss merckh |

12 JWA1, fol. 32v: 'so stand ab'.

This is again about the four positions and the displacements. Observe and pay attention that these are two positions too. The one who strikes by shooting over is also in a position. Here the positions and the four displacements come to an end; they all belong together, for there are many of them for the one who seeks them. Note the gloss.

This is about the travelling after. Learn it in two ways in the defence. Take two from it and your work begins with it. Check your encounters whether they are soft or hard; this you shall learn. The word *Indes* cuts severely. Travel after twice, in this way you can make the old cut with might. Be mindful. Note the gloss.

This is another travelling after, because you shall employ the travelling after on both sides. Make long steps if you want to stride, for travelling after has many forms. This is the travelling after: when someone has struck you and seeks to go away from you or pull away, then follow him and fight back. Note the gloss.

This is about running over. Note that the running over means that one runs over another one outwards with the pommel on the outside with force. This is an outer running over, and there are several more. For this reason, lift your sword up high when someone intends to run over you. In this way, he can't run over you. Note the gloss.

This is also about the running over. This is an inner running over. When you aim below and someone runs over you, you'll be harmed. When it clashes above, then set off, this I will praise. Make your work, softly or hard, and press it twice as you wish. Note the gloss.

28R Das ist auch ein vberlauffen vnd gewapp | enntt vnd got ein Ringen damit vnd darum | So merckh eben auff wan er wil in werffen | vber den glingken fuss vnd gehörtt zu dem | vberlauffen vil wen man suchen will | gloss merckh als es oben gmalt statt

28V Das ist auch ein vberlauffen mit dem | knopff vnd Ein vberlauffen gewappet | mit Ringen wie vor vnd Ruckh In vnd | wirff in damit vnd thu als es oben | gmalt statt gloss merckh also die | vberlauffen

29R Das ist von dem absetzen wer auff dich | hawt von dachab So setz im ab mitt | dem schwertt mit der kurtzen schneid sein | haw ist er waich so lass vmb dein kopff | gen vnd schlag in mit der zwerch ist | er aber hertt so schlag nit vnd nim | vnd verwend in an dem schwertt vnd[13] | gloss merckh[14]

29V **Ein absetzen**

Das ist ein absetzen auff der andern | Seytten vnd das nimptt man auch | mit der kurtzen schneid wie vor vnd | merckh auff die schwech vnd sterckh ob | er Sey waich oder hertt gloss merck

30R **Ein absetzenn**

Der ligtt in langen ortt

Das ist von dem absetzen haw stich schnelig | klichen letzen vnd der hinder ligt in dem | langen ortt vnd der vorder setzt im ab | mit ein absetzenn

30V **Ayn absetzen**

Ayn Lanng ortt

Das ist ain annder absetzen vnnd d*er* Voder setzt | dem ander den ortt ab gloss merckh etc

13 JWA1, fol. 34r: 'vnd arwait'.

14 JWA1, fol. 34r: 'merckh als es vnd[*en*] gmaltt statt'.

This is another running over which is armoured and has a useful wrestling. Therefore pay close attention when he seeks to throw the other one over the left foot. There's much that belongs to the running over which you shall try. Note the gloss as it is painted above.

This is another running over with the pommel and an armoured running over with wrestling as before. Wrench him to you and throw him with it. Do it as it is painted above. Note also the gloss about the running over.

This is about setting off. When he strikes at you from above, then set off his strike with the sword with the short edge. If he is soft then let go around your head and strike him with the *Zwerch*. But if he is hard, then don't strike but take and wind at his sword [and work]. Note the gloss [as it is painted below].

A setting off.

This is a setting off on the other side. You make it with the short edge as before. Pay attention to the weakness and the strength and whether he is soft or hard. Note the gloss.

A setting off.

He stands in the long point.

This is about setting off and how to hamper strikes and thrusts quickly. The posterior one stands in the long point while the anterior one sets off with a setting off.

A setting off.

A long point.

This is another setting off. The anterior one sets off the other one's point. Note the gloss.

31R **Ein absetzen**

Eeyn Lang ortt

Das ist wer will absetzen Hew stich schneliklichen | letzen vnnd wer auff[15] sticht wil so lug das dein | ortt drift vnd dem sein bricht von baiden seitten | drift wilttu schreitt*en* das ist auch ein ab | Setzen gloss merckh

31V **Ein absetzen**

Ein langer ortt

Das ist wer wil absetzen haw stich krum | schneligklichen letzen vnd ist auch Ein | absetzen gloss merckh Als Es oben gmalt | Statt

32R Das ist der erst oberschnidt vber die arm | vnd schneid ab die hört-tung schnel vnd flux | mitt deim gfertt vnd hab acht vnd nim | den schnidt mit macht vnd also thu gloss | merckh als es ~~vnden~~ oben[16] gmalt statt

32V Das ist der ander oberschnidt den nim auch | mit der langen schneid gloss merckh

33R Das ist von dem vndern schnidt der dritt | da schneid ab die hörttung durch baid arm | das merckh Als es oben[17] gmaltt statt vnd | nim das schnell vnd bald hinweg gloss | merckh

33V Das ist der viertt schnidt den nim auch wie | du in vnden gmaltt vinst vnd nim in | Starck oder waich wie du wiltt vnd flux | hinweg bald vnd bhend das dir dein schnidt | volendt gloss merckh

---

15 JWA1, fol. 36r: 'auf dich'.
16 JWA1, fol. 37r: 'vnden'.
17 JWA1, fol. 27r: 'vnden'.

A setting off.
A long point.

This is about setting off and impairing strikes and thrusts quickly. When someone thrusts onto [you], then watch out that your point hits and breaks his. Hit from both sides if you want to stride. This is another setting off. Note the gloss.

A setting off.
A long point.

This is about how to set off strikes and thrusts and impair them crookedly and quickly. This is another setting off. Note the gloss and as it is painted above.

This is the first upper cut over the arms. Cut off the hard ones quickly and swiftly with the encounter. Take care and make a cut with might. Do it like this. Note the gloss and as it is painted above.

This is another upper cut. Make it also with the long edge. Note the gloss.

This is about the third lower cut. Cut off the hard ones through both arms. Note this as it is painted above. Take it quickly and instantly away. Note the gloss.

This is the fourth cut. Make it in the same way as you find it painted below. Make it strongly or softly as you like. Take it swiftly away, quickly and nimbly, which completes the cut for you. Note the gloss.

34R Das ist auch ain oberschnidt den nim als du | in oben gmaltt sichst vnd schneid im ab die | herttung schnell vnd flux vnd hinweg das er | dich nit erfolg mit dem nachraisen gloss | merckh

34V Item das ist der Altt schnidt den treibtt der vorder | den nim also vnd zewch Im vnden auff durch | den Rechten arm starckh vnd langk vbersich | vnd fluchs hinweg gloss merckh

35R Das ist der ~~altt~~ ander altt schnidt zu der | andern Seitt*en* den nim also wie er oben[18] | gmalt statt vnd nim in flux auff vnd | hindersich hinweg das er dich nitt erlang | gloss merckh

35V Das ist von dem hend trugkhen das schwertt | druckh in die hend vnd thue das bhend dein | schwertt wend zu flech durch die hennd | gloss merckh[19]

36R Das ist auch ein hend trucken das nim also | vnd merckh nim dein schwertt vnd schlag | Im nach der blöss vnd truckh In von dir | starckh gloss merckh darauff vnd nim | war wie er gebar

36V Das ist von dem hengen etc zway hengen wend | aus ainer hand von der andern In allem gfertt | haw stich leger waich oder hertt mit seim gefertt | das ist das hengen das sich zwerchtt als du | es oben[20] gmalt sichst gloss merckh das ist | ein hengen auff der einen Seitten das ver- | -stannd Eben

37R Item das ist ain anders hengen auff der andern | Seitten das thutt er mit dem kurtzen schwertt | vnd heb den spitz bas vndersich wie du es | oben gmaltt Sichst vnd sind die zway heng*en* | zwu gutt versatzung zu allen sachen gloss | merckh

18 JWA1, fol. 29r: 'vnden'.

19 'gloss merckh' not in JWA1, fol. 29v.

20 JWA1, fol. 20v: 'vnden'.

This is another upper cut. Make it as you can see painted above. Cut off his hard ones quickly and swiftly away, so that he can't follow you with the travelling after. Note the gloss.

Item: It is the old cut that the anterior one does. Make it like this: Pull up below through his right arm, strongly and long upwards and swiftly away. Note the gloss.

This is the second old cut to the other side. Make it like it is painted above: Make it swiftly upwards and behind away so that he can't reach you. Note the gloss.

This is about pressing the hands. Press the sword into the hands and do this nimbly. Turn your sword to the flat through the hands. Note the gloss.

This is another pressing the hands. Make it like this: note to take your sword and to strike to his opening. Press him away from you strongly. Note the gloss for this and observe how he moves.

This is about the hanging. Turn to hangings out of one hand from the other in every encounter. Strike, thrust, position, soft or hard with his action. This is the hanging which uses the *Zwerch* as you can see in the painting above. Note the gloss. This is one hanging on one side. Understand this well.

Item: This is another hanging on the other side which he makes with the short sword. Lift the point strongly downwards as you see painted above. There are two hangings for good displacements in every matter. Note the gloss.

37v **Der brechfenster**

Item das ist das sprechfenster brich vnd mach | stand frelich vnd bsich Sein sach dein zwerch | mit der sterckh vnd wind Im vnden | durch vnd merckh als du es oben gmaltt | Sichst gloss merckh

38r **Ein langer Ortt**

Das ist von den vier valschen örttern vnd | ist des einen ein langer ortt vnd sy gend | zu baiden seitten da merckh auff das im der | lang ortt nim das gsicht gatt vnd ist gutt | zu allen sachen ein krieg damit ze machen gloss | merckh als du es oben gmaltt vindst

38v Das ist der ander verborgen ortt den nim | als du in oben gmaltt sichst der gett auch | zu dem gsicht darumb so merckh auff vnd | such was gutt Sey darauff

39r Das ist der tritt verborgen ortt vnd ist ein | valscher ortt den merckh als er oben gmaltt | Statt du macgst[21] auch nemen das durchgen | oder durchwechsel wie du wiltt das durch | gen nim also gee im mit der spitzen vnd*en* | durch gloss merckh

39v Das ist der viertt valsch ortt den nim auch | also du in oben gmaltt sihst vnd er dringt | dem das schwertt von der hend vnd ist auch | ain gutt*er* ortt vnd nim war vnd | merckh darauff Als du es oben gmaltt | sichst

40r Das ist aber ein valsch*er* ortt vnd ein valsch*er* | dritt den nim als du in oben gmalt sichst | vnd wan er auff dein Schwertt veltt | so wind auff mit deim schwertt so stastu | in der langen versatzung gloss merckh

21 The scribe probably intended to write 'machs' in the first place but changed his mind – without deleting the remaining orphaned letter 'c'.

The *Brechfenster*.

Item: This is how to make and break the *Sprechfenster*. Stand unconcernedly and watch what he intends to do. Make your *Zwerch* with strength and wind through below. Notice this as you can see painted above. Note the gloss.

A long point.

This is about the four deceitful points. This, for one thing, is a long point which can be employed on both sides. Pay careful attention that the long point is directed into his face, which is good for everything and useful for the *Krieg*. Note the gloss and as you can see it painted above.

This is the second secret point. Make it as you see it painted above. It goes to the face too. Therefore be careful and pay attention to what serves you well with this.

This is the third secret point, and it is a deceitful point. Notice how it is painted above. You can also choose to go through or change through as you like. Go through like this: go through with the tip below. Note the gloss.

This is the fourth deceitful point. Make it in the same way as you can see in the painting above. It presses his sword out of the hands. It is a useful point which you shall observe and pay attention to as you see it painted above.

This is yet another deceitful point and a deceitful step which you shall make as it is painted above. When he falls on your sword, then wind up with your sword. In this way you stand in the long displacement. Note the gloss.

40V **Der statt in einhorn**

Das ist auch ein valscher ortt vnd gett | aus dem sprechfenster vnd den nim | als du es oben gmaltt sichst vnd ist | das einhorn gloss merckh

41R Item das ist die kron der sein schwertt gewappent heltt | in der hand schon vnd nim war schon mit der | kron die ist gutt In alle weg zu allen weren vnd | bis starckh darin oder schwach wie du wiltt gloss | merckh

41V Das ist der schnidt durch die kron vnd den | nim also schon als du es gmaltt sichst in | allen sachen wilttu ein schnitt machen | gloss merckh schon auff den schnidt in der | kron

42R Item das ist der bruch auff die kron da merckh also | schon als du es gmaltt sichst vnd der das schwertt | vbersich hebtt der bricht die kron mitt dem | Reissen ain schwertt mit dem krutz da lug auff | wen er Reist im die Kron hinweg gloss merckh

Hie hatt der lang Zettell ein End das helff | vns gott also behend vnd ist die kron das | Aller letst vnd ist das aller best etc

1523

He stands in the *Einhorn*.

This is another deceitful point and can be made from the *Sprechfenster*. Make it as you can see painted above. This is the *Einhorn*. Note the gloss.

Item: This is the *Kron*, shown by the one holding the sword with his hands in the armoured position. Pay close attention to the *Kron*, for it is good for defending in every way. Use it strongly or weakly, however you like. Note the gloss.

This is the cut through the *Kron*. Make it promptly, as you see it painted above, for everything when you want to cut. Note the gloss diligently for the cut through the *Kron*.

Item: This is the break against the *Kron*. Pay careful attention as you can see it painted above. He who lifts the swords upwards breaks the *Kron* by wrenching the sword with the hilt. Watch when he wrenches his *Kron* away. Note the gloss.

Here the long Notes come to an end. God help us swiftly. The *Kron* is the very last and the very best.

1523

Johannes Liechtenauer

LONG SWORD

VERSES

43R

Ein gutt gemain ler des langen schwertz

Wiltu kunst schawen so bis lingks vnd rechts | hawen vnd lingk mit Rechtem ist das du starckh | begerest zu vechten wer nach geitt Haw-en der | darff sich kunst wenig frawen haw im nachentt | nach was du wiltt kain durchwechsel kumpt an | deinen schiltt zu kopff zu leib vicht nit vermeid | mit gantzem leib vicht was du starckh begerst | ze-treiben hernach was du schlechst vicht nitt oben | lingks so du Rechts bist vnd ob du glingks bist im | Rechten auch ser hinckst vor vnd nach die zwai | ding send aller kunst ein vrsprung schwech vnd | sterckh Indes wartt dein arwaitt damit merckh | so magstu lernen mit kunst arwaitten vnd | wer erschrichkt gern kain vechten sol er lernen[1] | funf hew lernen von der Rechten hand gerne wider | die gewere wir glau-ben zu kunsten gern zu | kumen

Das ist der text vnd die auslegung | der schaittler

Zornhaw krumphaw zwerch durch haw schilcher nach | raisen vber-lauffen haw setzt durch wechsel auff | alber abschnidt hend druckhen heng winden | mit blosem schwertt schlachen streich stich mitt | stös-sen

Das ist die ausrichtung vom zornhaw

Wiltu merckhen wer dir oberhawtt zornortt im | drott herwider wind stich sicht ers so nim es wider | das eben merckh haw stich leger waich oder hörtt | Indes vor vnd nach dem krieg sey dir nit gach | wer des kriegs vnden bremett der wirtt oben | bschemett In allen winden haw vnd ~~sch~~ stich

1 JWA1, fol. 42r: 'vechten er nimer sol lernen'.

A good common lesson of the long sword

If you wish to observe the art, be left and strike right; and rightfully left means that you desire to fence strongly. Who goes after strikes shall not take much pleasure in the art. Strike closely after him whatever you wish. No changing through will come at your shield. Do not avoid fighting towards the head and the body. Fight with the entire body whatever you want to do strongly. Hereafter: whatever you strike, do not fight above on the left if you are right. And if you are left, you are much hampered on the right. Before and after, these two things are the origin of the entire art. Weakness and strength. Mind the *Indes*, make your work with it. So you may learn to work skilfully. Whoever is easily startled shall never learn how to fence. Study well five strikes from the right side against the weapon, so we believe to make good use of the art.

The text and the explanation of the *Schaitler*[2]

Zornhaw, *Krumphaw*, strike *Zwerch* through, *Schilcher*, travelling after, running over, place strikes, change through – onto *Alber*, slice off, press hands, hang, wind, strike with the bare sword, beat, thrust with blows.

The explanation of the *Zornhaw*

You want to know this: whoever strikes an upper strike to you, the *Zornort* threatens him. Wind against, thrust. If he sees it, take it again. Remember this well: strike, thrust, position, soft or hard, *Indes*, before and after. Don't be hasty in the *Krieg*. Who desires the *Krieg* below will be put to shame above. Find in every winding strikes, thrusts and cuts.

---

2 In this place, something definitely got mixed up. The *Scheitler* is only one of the five secret strikes, the remaining four of which are listed in the following paragraph. Other Liechtenauer-related sources have no or an entirely different headline here. The explanation of what the *Scheitler* is and how it is performed comes much later, so the scribe has either misread this sequence or he copied it erroneously from another spot in his notes.

43v Schneide finden auch soltt du nit meiden hau stich | schneide zuckh in alem treffen wiltu den maister | Effen etc

Das ist von den vier blössen

Die vier blöss wis zu Remen so schlechst gwiss | on alle far on alle zweiuel wie er gebar hastu es | vernomen zu kaim schlag mag er ko-men etc

Das ist wie man die vier blöss brechen sol

Wiltu dich Rechen die vier blöss kunstlichen brechen | du oben dup-plier vnden zu Recht muttier ich sags | dir virwar sich bschitzt kain man one far hastu | vernomen zu schlagen mag ir kainer komen das | han ich vernomen

Das ist die ausrichtung vom krumphaw

Krumphaw auff behend wirff den ortt auff die | hend wer wol versetzt mit schreitten er vil hew letzt | haw krumb zu der vlechen wilttu den maister schwechen | wan es glitzt oben so setz ab das wil ich loben krum | vnd schlecht vnd kurtzhaw dein durch wechsel damit | schaw haw krumb wer dich Irrett der edel krieg | in verwirett das er nit waist virwar[3] wie er | gebar etc

Die ausrichtung vom zwerchhaw

Die zwerchhaw benimptt was vom tag her kompt | die zwerch mit der sterckhe dein arwait damitt | mercke ochs pflug alber vom tag ist wol sein fug | was sich wol zwercht mit springen deim [!] haubt | geferet was veler verfiert nitt[4] vnden nach | wunsch herrierentt verkerer zwinger duchlauffer | damit Ringett nim dein elenpogen gwiss

3 JWA1, fol. 42v: 'wo er ist on alle far'.

4 Unclear. The 't' in 'verfiert' is not unambiguously legible. Also, the letter 'i' is missing in the word 'nitt' although a corresponding tittle exists. 'nitt' is completely absent from JWA1.

Also you shall not avoid strikes, thrusts and cuts. Twitch in every encounter if you want to fool a master.

About the four openings

You shall know how to attack the four openings, so you will strike surely without any danger and doubt, however he moves. If you have understood this, he may not perform any strike.

How to break the four openings

If you want to take revenge, break the four openings skillfully: Above, use *Duplieren*, and below rightfully *Mutieren*. I tell you indeed that no one can protect himself without danger. If you have understood this, nobody may be able to strike. This I have understood.

The explanation of the *Krumphaw*

Make the *Krumphaw* upwards, throw the point on the hands. Who can displace well, can impair many strikes with stepping. Strike the *Krump* to the flat if you want to weaken a master. When it clashes above, set off, which I wish to praise. Strike in a crooked and straight and short way. Meanwhile, take care of your changing through. Strike the *Krump* to those who want to distract you, the noble *Krieg* will confuse them so that they don't know [where they may be without danger and] how to act at all.

The explanation of the *Zwerchhaw*

The *Zwerchhaw* takes what comes from above – *vom Tag*. Remember to make the *Zwerch* with strength in all your work. *Ochs*, *Pflug*, *Alber*, *vom Tag* are just right. It is good to jump with a *Zwerchhaw*, it is a danger to the head. The *Fehler* does not mislead below; it touches as you like. *Verkehrer* enforces running through; wrestle at the same time. Take the elbow certainly.

44R Spring im wol in die wage nim den glingken ellen- | pogen zwifach mit macht virbas schreitt in lincken | vnd bis nit lass

Die ausrichtung vom schilthaw[5]

Schilcher ein bricht was buffel schlecht oder sticht wer | wechsel treibt mit schilcher wirtt er daraus beraubt | schilcht er dich kurtz an dein durch wechsel gsigt Im | an schilch zu dem ortt vnd nim den halss on vorcht | schilch zu dem haubt oben wiltu in betoben etc

Die ausrichtung von dem schaitelhaw

Der schaitler dem antlitz ist gefar mit seiner kar | der brust vast virwar was von im kumpt dar die | kron das abnimpt schneid durch die kron so brichstu sy | hartt vnd schon die straich druckh durch mitt schneiden | Sy abzuckh

Die ausrichtung von den vier legern

Die vier leger allain daruon heltt man vnd | fleuchts die gemain ochs pflug alber vom tag ist | wol sein fug

Die ausrichtung von den vier versetzen

Vier sind der versetzen vnd die leger auch serr letzen | doch vor versetzen hiett dich gschicht es dir nott es | miet dich ob dir versetzt ist vnd wie das darkumen | ist her was ich dir Ratt kum frie vnd nitt zu spatt | streich auff haw schnel mit dritten selb an vier | enden vnd bleib darauff so lernestu Enden

5 Most likely a scribal error which also appears in JWA1, fol. 42v.

Jump well into the scales and take the left elbow twice with force. Step forwards to the left and don't be lax.

The explanation of the *Schilthaw*

Schilcher breaks into what a buffalo strikes or thrusts. He who makes the changer is spoilt with it by the *Schilcher*. If he makes the *Schilcher* short against you, your changing through will give you victory. Squint – or make the *Schilcher* – towards the point and take the neck without fear. Squint – or make the *Schilcher* – towards the head above if you want to daze him.

The explanation of the *Schaitelhaw*

The *Schaitler* is a danger to the face and, with its turn, very much to the breast. Whatever comes from it takes the *Kron*. Slice through the *Kron*, so you break it hard and readily. Press through the strikes; twitch them away with slicing.

The explanation of the four positions

Only hold to four positions and back away from the common ones. *Ochs*, *Pflug*, *Alber*, *vom Tag* are the best opportunities.

The explanation of the four displacements

There are four displacements that hurt the positions very much. But beware of displacing; if it happens, it gives you distress. It troubles you if you have been displaced, however it has come to pass. Listen to my advice: come early and not too late, strike upward, make sure to hit quickly with steps on four ends. Stay on them, so you learn how to make an end.

Von dem nachRaisenn

Nachraisen lere zwifach schneidett in der were | zway darauss nim dein arwait darauss begine | vnd brieff sein gfert sein leger waich oder hertt das | selbig lere Indes das wort schniedett sere nach- | raisen zwiuach nim du den altten schnidt mit macht

44v **Die ausrichtung von dem vberlauffen**

Der vnden Remet vnd vberlauft der wirt oben | gschemett wan es glitzt oben so setz ab das wil ich | loben dein arwait mach waich oder hertt vnd | druckh das zwiuach

Die ausrichtung von dem absetzen

Lere absetzen haw stich krum schneligklichen letzen | wer auff dich sticht so lug das dein ortt drift | vnd dem seinen bricht von baiden seitten driff alle | mal wilttu schreitten

Von dem durchwechsell

Der durchwechsel lere von baiden seitten stich mitt | Sere vnd wer auff dich sticht byndett[6] durchwechsel | in schier windett

Die ausrichtung von dem treffen vnd zucken

Dritt no in[7] die binden Zucken geitt gutte vinden | drift er so zuckh mer zuckh aus allen treffen | wlttu [!] den maister Effen

Die ausrichtung von dem durchlauffen

Durchlauff las hengen mit dem knopff vnd greiff | wiltu Ringen vnd wer gegen dir streicht sterckhe durch- | lauff damit merckhe

6 'n' added above the 'y'.

7 Unclear; this could also be read as 'nein', but that doesn't make much sense.

About travelling after

Learn travelling after in two ways, it cuts into the weapon. Take two of them and start to work from there. Consider his passage of arms and his positions, whether they are soft or hard. This you must learn: The word *Indes* cuts strongly. Travel after twice, make the old cut with force.

The explanation of running over

Who aims below and runs over will be put to shame above. When it clashes above, set off; this I will praise. Make your work soft or hard and press it twice.

The explanation of setting off

Learn to set off and to impair strikes and thrusts crookedly. When he thrusts to you, pay attention that you hit with your point and break his from both sides. Hit every time you want to step.

About changing through

Learn to change through from both sides. Thrust with injury. Whoever thrusts and binds to you, the changing through finds him uncovered.

The explanation of hitting and twitching

Step close when you bind. Twitching gives good finds. If he hits, twitch further. Twitch from all encounters if you want to fool a master.

The explanation of running through

Run through. Hang with the pommel and grasp if you want to wrestle. Be strong against him who strikes against you. Remember to run through in the course of this.

Die ausrichtung von dem abschneiden

Schneidt ab die hertte von vnden yn baiden gfertte | vier send der schnidt zwen vnden zwen oben mitt

Von dem hend truckhen

Das schwertt bind zu der flöche truckh in die | hend von dir dein schneid wend zu der flech durch | die hend

Von den Zwaien hengen die ausrichtung

Zway hengen werden auff einer hand von der andern | in allem gfertt haw stich leger waich oder hörtt In | irem gfertt etc

45R **Von dem sprechfenster die ausrichtung**

Das sprechfenster machen stee frolich vnd sich sein | sache wer sich vor dir ziechett abe schlachen das er | schnabe thutt er sich gegen dir greisen [!] schlagen das | er seise

Das ist die ausrichtung vnd die bschliesung der gantzen | kunst wer wol sicht vnd endlich bricht es bricht | ettlicher in dreien vinden wer Recht wol hengett | vnd windett da mit es bringett wachett mit | rechtem betrachten vnd einer der da windett | vnd 23[8] sind der winden vnd vierzal mitt | versatzung von baiden seitten acht winden verr | mit schreitten tritt vnd brieff das gfertte ob sy | Sind waich oder hertte also endett sich maister | liechtnauers kunst des langen schwertts

8 Interesting scribal error that appears consistently throughout the Wilhalm group of manuscripts. Every other source has the – arithmetically correct – sum of 24 windings.

The explanation of cutting off

Cut off the hard ones from below in both encounters. There are four cuts: two below and also two above.

About pressing the hands

Bind the sword to the flat, press into the hands away from you. Turn your edge to the flat through the hands.

The explanation of the two hangings

There are two hangings on one hand from the other in every encounter. Strike, thrust, position, soft or hard in their encounters.

The explanation of the *Sprechfenster*

How to make the *Sprechfenster*: Stand unconcernedly and watch his affairs. When he draws back in front of you, strike so that he surrenders. If he wants to grab you, strike him so that he dashes away.

This is the explanation and the conclusion of the entire art

Who observes well will finally break. There are many who break in three findings. Who hangs quite well will accomplish windings at the same time. Keep watch, observing properly one who winds. There are 23 windings and a number of four with displacements from both sides: eight windings. Go far with steps and test the encounters whether they are soft or hard. Hereby ends Master Liechtenauer's art of the long sword.

Oberhaw zornhaw zornortt der krieghaw die vier | blöss krumphaw zwerchhaw ochs vnd pflug | schilcher haw schaitlerhaw die vier leger die | vier versetzen das nach Raisen vberlauffen | absetzen abschneiden zway hengen das sprech | fenst*er* die kron die stuckh des schwerts[9] Johannes | liecht*e*nawers

Die siben maist*er* haw

Der zornhaw der krumphaw der halbhaw | der kreizhaw der zwerchwechselhaw der | schaittelhaw der kronhaw etc

1523

9 JWA1, fol. 44r: 'des langen Schwertz.'

Oberhaw, *Zornhaw*, *Zornort*, the *Krieghaw*, the four openings, *Krumphaw*, *Zwerchhaw*, *Ochs* and *Pflug*, *Schilcherhaw*, the four positions, the four displacements, the travelling after, running over, setting off, cutting off, two hangings, the *Sprechfenster*, the *Kron*: the pieces of the [long] sword by Johannes Liechtenauer.

The seven master strikes

The *Zornhaw*, the *Krumphaw*, the *Halbhaw*, the *Kreuzhaw*, the *Zwerchwechselhaw*, the *Schaitelhaw*, the *Kronhaw*.

1523

Johannes Liechtenauer

LONG SWORD

VERSES & GLOSSES

45v **Das ist der text von den dreien winden das ist ein haw | ein stich ein schnit ein[1] streich vnd ist auch der | text von dem langen zettel des langen schwertts | des liechtnauers kunst vnd ausrichtung gloss**

Item hie merckh wie du aus den zwaien obern hengen | soltt treiben das ist aus dem ochsen soltu treiben Im | winden von baiden seitten von der Rechten seitt*en* 2 von | der lingk*en* seiten 2 vnd treib die also wen du mit dem | Rechten zu im kumst so stand vor mit der Rechten | Seitten in den ochsen hawett er dir den oben ein zu | deiner lingken Seitt*en* so wind gegen seinem haw die | kurtz schneid an sein schwertt aus dem ochsen vnd | stich im oben ein zu dem gsicht das ist ein winden | versetzt er den stich so bleib am schwertt vnd wind | wider auff auf der Recht*en* seitt*en* aber in den ochsen | die kurtz schneid an sein schwertt vnd stich nit[2] oben | ein zu dem gsicht das sind zwai winden am schwert | aus dem obern hengen vor der Recht*en* seiten Item | das ander ober hengen treib also wen du mit dem | zuvechten kumst so stand von deiner lingke seitten | in dem ochsen hawtt er dir den oben zu deiner Rechte | Seitten so wind gegen seinem haw die lang schneid | an sein schwertt vnd stich im oben ein zu dem gsicht | das ist ein winden setzt er den stich ab so bleib an dein*em* | schwertt vnd wind wider auff die lingk seitten | in dem ochsen die lang schneid an sein schwertt vnd | im ochsen oben ein zu dem gsicht das sind die vier | winden aus den obern zwaien hengen Item nun | solttu wissen das aus dem pflug von baiden seitten | das sind die zwai ~~hengen~~ vnder hengen auch soltu | treiben vier winden mit allem Irem gfertte als | aus dem obern hengen also werden der winden | Achte vnd merckh als oft du windest so ge-

46r denckh in ein Jettlich winden besunder an den haw | vnd an den stich vnd an die schnidt also kumestu aus | dem achtenden winden 24 vnd wie du hew vnd | stich schnidt treiben soltt das vindestu ales dar-

1 The word is almost illegible. (Water damage?)

2 Recte: 'mit'. The same flaw appears in JWA1, fol. 44v.

This is the text of the three windings; which is a strike, a thrust, a cut and a hit. It's also the text of the long Notes of Liechtenauer's art and explanation of the long sword. The gloss.

Item: Note here what to make out of the two upper hangings, which is the *Ochs*. You shall make them while winding from both sides: two from the right side, two from the left. Make them this way: when you come to him with the right, stand forward with the right side in the *Ochs*. If he strikes above towards your left side, wind from the *Ochs* against his strike, the short edge at his sword. Thrust above towards his face; this is one winding. If he displaces the thrust, remain at the sword and wind again to the right side, again into the *Ochs*, with the short edge at his sword. At the same time, thrust above towards the face. These are two windings at the sword from the upper hanging on the right side.

Item: Make the other upper hanging like this: when you come to him in the *Zufechten*, stand in the *Ochs* on your left side. If he strikes above to your right side, wind against his strike, with the long edge on his sword. Thrust above to his face. This is one winding. If he displaces the thrust, remain at your sword[3] and wind again to the left side into the *Ochs*, the long edge on his sword; and in the *Ochs* above towards the face. These are four windings from the two upper hangings.

Item. Now you shall know that that there are two lower hangings from the *Pflug* on both sides. From these, you shall make four windings with all their encounters as from the upper hangings, so there are eight windings. Remember that whenever you wind, you shall consider in each single winding particularly the strike, the thrust and the cut. In this way, you get 24 out of the eight windings. How to execute the strikes, thrusts and cuts you can find written in the pieces ahead.

3 Other sources in the tradition lines of Master Liechtenauer say 'remain at *the* sword', refering to the opponent's weapon, which in fact makes much more sense.

uon | in den stuckhen geschriben auch solttu vor allen | dingen die 8 winden verttigklich*en* treiben lernen | mit schreitten zu baiden seitten vnd merckh | als bald du windest so soltu in einem Jegklich*en* | winden besunder brieffen nit mer den die zwai | gfertt ob sy sind waich oder hertt am schwert ist | darnach treib die stuckh als sy dir den vor in dem | negsten vorgeklertt sind worden

Das ist der text vom dem schilch*er*

Lere wen du mit dem zu vechten zu gangest so | soltu schilchen mit dem gs~~ch~~icht ober kurtz gegen dir | Sicht[4] das solttu bei dem erkenen wen er dir zu hawt | Recht Raicht er den die arm mit dem schwertt haw | nit lang von Im so ist das schwertt vor im kurtz | leistu dan vor im in der hutt aber wil er dir mit | dem schwertt drauff vallen so ist es aber verkurtzt | legt er sich den gegen dir in die hutt des ochsens od*er* | des pflugs so ist es aber verkurtzt auch wis das alle | winden mit dem schwertt vor dem man kurtz send | vnd verkurtzen das schwertt vnd alle die also gegen | dir vechten den wechsel frölich durch aus hawen vnd | durch aus stechen mit dem langen ortt damit zwing- | estu sy das si das miessen versetzen vnd sich schlach*en* | lassen schilch zu dem ortt vnd nim den hals one | vorcht gloss merckh das ist ein bruch wider | den langen ortt mit einer betriebtnus des gsichts

46v den treib also wen du mit dem zuvechten komst statt | er den vnd helt den ortt gegen dem gsicht oder brust | so haltt das schwertt in dem Rechten ochsen vnd schilch | mit dem gsycht zu dem ortt vnd thu als du in darzu | hawen wellest vnd haw starckh aus dem schilch*er* | mit der kurtzen schneid an sein schwertt vnd scheub | im den ortt damit langk ein zu dem hals mit eim | tritt des Rechten fus

4 JWA1, fol. 45r: 'ficht' – which makes more sense in the context.

Above all, you shall learn how to execute the eight windings nimbly, with stepping to both sides. Note: As soon as you wind, you shall particularly check in each winding nothing else but whether they are soft or hard at the sword in these two encounters. Make the pieces accordingly as they are explained in the next section.

The text about the *Schilcher*

Learn this: When you go to him with the *Zufechten*, you shall squint with the eyes to see whether he looks[5] shortly at you. You can recognise that by this: When he strikes at you and he does not stretch out his arms long, then the sword before him is short.

If you stand before him in the guard but he strives to fall onto it with the sword, then he is shortened. If he stands against you in the guard *Ochs* or *Pflug*, he is also shortened. You shall also know that all windings with the sword in front of the opponent are short and they shorten the sword.

All these who fight against you in this way, you shall change through unconcernedly from strikes and from thrusts with the long point. With this, you force them into displacing and getting hit.

Squint to the point and take the throat without fear.[6] Note the gloss: this is a break against the long point with a deception of the eyes.

Make it like this: when you come to him with the *Zufechten* and he stands against you, holding the point against your face or breast, then hold your sword in the right *Ochs* and squint with the eyes to the point and pretend to strike there. Strike strongly from the

---

5 Considering that the original probably should have been 'ficht' instead of 'sicht', the passage would read: 'whether he fights shortly against you'.

6 Master Liechtenauer's verses, notoriously difficult to understand as they are, have been heavily modified by Jörg Wilhalm and are set here in bold to help the reader identify them more reliably. See the editorial guidelines, p. 51.

das ist der text schilch zu dem | obern haubt behend wilttu bedreben gloss merckh | das ist ain ander bruch wen er gegen dir stet in dem | langen ortt wilttu in auff die hend schlachen so | schilch im mit dem gsicht zu dem haubt vnd thu | als du in darauff schlachen wellest vnd schlag | in aus dem schilch*er* mit dem ortt auff die hend

Das ist der text von dem schaitelhaw

Der schaitler dem antlitz ist gefar mit seiner kar | der brust vast virwar[7] was von im kumpt die | kron das abnimpt der schnidt durch die kron so | brichstu sy hartt vnd schon die starckh druckh | mit schneiden sy abzuck glosa dem schaitler bricht | die hutt alber vnd ist dem antlitz gefar vnd der | brust mit seiner kar gar geferlich den treib also wen | du mit dem zuvechten kumst legt er sich gegen dir | in die alber so setz den lingken fus vor vnd haltt | dein schwertt mit aufgereckten armen hoch vber | dein haubt in der hutt von tag vnd spring zu | im vnd haw mit dem langen schwertt schneid starckh | von oben nider vnd bleib mit den armen hoch vnd | das ein ghultz an seinem [!] schwertt baid vbersich sten | das haist die kron vnd vertt damit auff vnd stost | dir den ortt so wend dein schwertt vnd sein kron | durch mit dem schnidt vnder dem knopf zu seinem | arm vnd truckh also ist die kron wid*er* gebrochen
47R vnd mit dem nider truckhen schneid vnd zeuch dich | damit ab

Das ist der text von den leger oder hutten

Vier leger allain daruon haltt vnd fleuch die gemain | ochs vnd pflug alber vom tag sei dir nit vnmer gloss | merckh die vier leger das sind die vier hutten do du | aus vechten solt die erst hutt haist der ochs da schick | dich also mit stee mit dem lingken fuss for vnd | haltt dein schwertt neben seiner Rechten seitten mit | dem ghultz vor dem

7 JWA1, fol. 45v: 'gefar.'

Schilcher with the short edge at his sword and with a step of the right foot, shove your point long towards his throat.

The text: **Squint nimbly to the upper head if you want to bring about his undoing.** Note the gloss: This is another break. If he stands against you in the long point and you want to strike over his hands, then squint with your eyes to his head and pretend to strike onto it. Strike from the *Schilcher* with the point onto his hands.

The text about the *Schaitelhaw*

The *Schaitler* is a danger to the face and with its turn, indeed severely to the breast. What comes of it, the *Kron* takes away. With a cut through the *Kron*, you break it hard and fairly. Press the strength, twitch away with cuts. Gloss: the *Schaitler* breaks the guard *Alber* and is a danger for the face, and with its turn, it is dangerous to the breast. Make it like this: when you come to him with the *Zufechten* and he positions himself in the *Alber*, then put your left foot forward and hold your sword with raised arms high above your head in the guard *vom Tag*. Jump to him and strike with the long sword edge strongly down from above. Remain with your arms high, so that one quillon on your[8] sword is directed upwards. This is called the *Kron*. If he goes up and pushes your point, then turn your sword through under his *Kron* with a cut under the pommel towards his arm and push. In this way, the *Kron* is broken again. Cut while pressing down, and retreat at the same time.

The text of the positions or the guards

Think only of the four positions and stand back from the common ones. *Ochs*, *Pflug*, *Alber*, *vom Tag* shall not be unknown to you. Note the gloss: The four positions are the four guards from which you shall fence. The first guard is called the *Ochs* and you shall make it in this way: Stand with your left foot forward and hold your sword

8 Not 'his', as the manuscript suggests.

haupt das die kurtz schneid gegen | dir stand vnd im den ortt gegen dem gsicht Item | zu deiner lingken seitten schick dich also in dem ochsen | stand mit dem Rechten fus vor vnd haltt dein schwertt | neben deiner lingken seitten mit dem ghultz vor dem | haubtt das die lang schneid gegen dir stee vnd im | mit dem ortt gegen dem gsicht das ist der ochs | zu baiden seitten

Item die ander hutt die haist der pflug so schick dich also | mit setz den lingken fuss vor vnd haltt dein schwertt | mit verkertten henden mit dem knopff vndersich | neben der Rechten seitten zu der huff vnd das die | kurtz schneid oben sey vnd der ortt gegen dem gsicht | It*e*m zu der lingken seitten schick dich also in den pflug setz | den Rechten fus vor vnd haltt das schwertt neben | der lingken seitten mit dem knopff vndersich der | huff vnd das die lang schneid oben sey vnd der ortt | im gegen dem gsicht das ist der pflug zu baid*e*n | Seitten

Item die dritt haist der alber da schick dich also mit | setz den Rechten fus vor vnd halt dein schwert | mit ~~auffgereckten armen~~ dem Rechten arm | vor dir mit dem ortt spitz auff der erden vnd das
47v die kurtz schneid oben gwand sey

Item die viertt hutt haist vom tag da schick dich | also mit vnd setz dein lingken fus vor vnd halt | dein schwertt midtt aufgereckten armen hoh | vber dein haubt mit dem spitz vnd stee also in | der hutt

Das ist der text von vier versetzen

Vier send der versetzen die die maist*e*r oder leger auch | sere letzen vor versetzen hiett dich gschicht ess dir | es miett dich gloss merckh du hast vör gehertt | von den hutten so soltu nun wissen was[9] hew | im versetzen die vier hutten brechen wiss das | die kron versetzen nit darzu gehertt wan es | Sind vier hew die sy brechen

9 Almost illegible word.

next to your[10] right side with the hilt before the head so that the short edge faces you while the point is directed to his face. Item: On your left side, make the *Ochs* as follows: stand with your right foot forward and hold your sword next to your left side with the hilt before the head so that the long edge faces you while the point is directed to his face. This is the *Ochs* on both sides.

Item: The second guard is called the *Pflug* and you make it like this: put your left foot forward and hold your sword with twisted hands with the pommel downwards next to your right side at the hip so that the short edge shows up and the point is against the face.

Item: On the left side, make the *Pflug* as follows: Put your right foot forward and hold the sword next to the left side at the hip with the pommel down so that the long edge shows up and the point is against his face. This is the *Pflug* on both sides.

Item: The third guard is called the *Alber* and you make it like this: put your right foot forward and hold your sword with the right arm before you, with the point on the ground and the short edge showing upwards.

Item: The fourth guard is called *vom Tag* and you make it as follows: put your left foot forward and hold your sword with raised arms high over your head with the point. Stand like this in the guard.

The text about four displacements

There are four displacements that harm the masters or the positions greatly. Beware of displacing, if it happens to you, it will trouble you. Note the gloss: You have heard before about the guards. Now you shall know which strikes break the four guards while displacing. Know that displacing the *Kron* does not belong to this because it is four strikes that break them.

10 Not 'his', as the manuscript suggests.

Item der Erst haw ist der krumphaw der bricht | die hutt aus dem ochsen Item der ander haw ist die | zwerch bricht die hutt vom tag Item der dritt | haw ist der schilcher der bricht die hutt die da haist | der pflug Item der viertt haw ist der schaitler der | bricht die hutt die da haist der alber vnd wie du | vier huttn mit den hewen brechen solt das vind- | estu daruor zu den hewen gschriben darum so | hiett dich das du nitt vil versetzest wilttu | anderst nit gschlagen werden

Das ist der text von dem versetzen

Ob Dir versetzt ist vnd wie das dar komen ist her was | ich dir Ratte Reis ab haw schnel mit dritten gloss | merckh ist das dir ainer versetzt hat vnd wil sich | vom schwertt nit abziechen vnd maint er wel | dich zu kainem stuckh lassen komen so thu als | ob du dich vom schwertt wellest ziehen vnd

48R zuckh dein schwertt an dich bis an das halb tail der | klingen vnd var damitt bald am schwertt auff vnd | haw im schnel mit dem dupplieren oder mit der | kurtzen schneid zu dem kopff gegen seinem ortt als ob | du ims welest nemen vnd bleib am schwertt vnd | haw im schlecht an die klingen mit der langen schneid | wider ein zu dem kopff

Das ist der text von den vier ansetzen

glos merckh[11] Setz an vier Enden bleib darauff lere willtu | Enden das ansetzen ist ein erst stuckh wan es gatt | nach den vier blössen vnd gehörtt zu treiben wen | du bald vnd wilt geben Item das ansetzen treib | also wen du mit dem schwertt mit dem vechten zu | im komst so leg dich mit dem schwertt in die hutt | des ochsens oder des pflugs wil er dir dan oben ein | hawen oder stechen vor seiner Rechten seitten so kom | vor mit der versatzung vnd scheub im den langen | ortt ein zu der plöse seiner lingke seitten vnd wartt | ob du im anmigest

11 Added on the right margin. In JWA1, fol. 46v this passage is written between the verse and the explanation, namely between 'wilttu Enden' and 'das ansetzen'.

Item: The first strike is the *Krumphaw* which breaks the guard of the *Ochs*. Item: The second strike is the *Zwerch* which breaks the guard *vom Tag*. Item: The third strike is the *Schilcher* which breaks the guard called the *Pflug*. Item: The fourth strike is the *Scheitler* which breaks the guard that is called the *Alber*. How you shall break the four guards with the strikes you'll find written further ahead in the strikes. So be careful not to displace too much, so you don't get hit.

The text about displacing

If you have been displaced, however it has come to pass, listen to my advice: tear yourself away, strike quickly with steps. Note the gloss. If it happens that someone has displaced you and doesn't want to retreat from the sword in order not to let you get to any pieces, then pretend to pull away from the sword. Twitch your sword towards you up to the middle of the blade and go immediately up at the sword and strike quickly with the *Duplieren* or with the short edge to his head against his point as if you wanted to take it away from him. Remain at the sword and strike plainly at the blade with the long edge, again to his head.

The text of the four ways of setting on

Set on at four points. Learn to stay on them if you want to end. Note the gloss: The setting on is a serious piece because it aims at the four openings and shall be executed when you want to make an end[12] quickly. Item: Make the setting on like this: when you come to him fighting with the sword, then position yourself with the sword in the guard of the *Ochs* or the *Pflug*. If he intends to strike to you above or to thrust before his right side, then come forward with a displacement and shove the long point into the opening of his left

12 The manuscript says 'vnd' (and) rather than 'end' here, thus corrupting the intended meaning.

setzen oder hawtt er dir oben ein | von seiner lingken seitten so scheub im den ortt ein zu | der blöss seiner Rechten seitten Item were er dir | zu hawt von vnden auff von seiner Rechten seitten | So scheub im den ortt ein vnd setz im an des selben | gleichen thu auch wen er von der lingken seitten | vnden auf hawt wertt er den das ansetzen so ver- | setz vnd bleib mit dem schwertt auff dem seinen | vnd arwait behendigklich zu dem negsten blöss | So lastu in zu kainer kunst ku-men

Das ist der text von dem Nachraisen

Nachraisen lere tzwiuach oder schneid in die were zway | auserer nym dein arwait darnach begine vnd | brieff die gfertt ob sy sind waich oder hertt

48v glos merckh der nachraisen ist vil vnd mangerlay vnd | geheren die zu treiben mit groser virsichtigkait | gegen den vechtern die da aus va-ren vnd in langen | hewen vechten vnd sunst von Rechter kunst des | Schwertts nit halber

Item das erst stuckh aus dem nachraisen treib also wan | du mit dem zu vechten kumst so setz den lingken | fus vor vnd stee in der hutt vom tag vnd sich gar eben | was er gegen dir vicht hawet er dir oben glingks ein | von seiner achsel das versetz im nitt vnd wartt das er | dich mit dem haw nit vberlang vnd truckh merckh | die weil sein schwertt mit dem haw vndersich gatt | gegen der erden so spring hin zu der blöss seiner | gerechten seitten Ee das er mit seinem schwertt | auf kumpt so ist er gschlagen

Das stuckh haist die Eisnin wine portt

Item wan er sich verhawt vnd du im nach hawest fert | er den bald auff mit dem schwertt vnd versetzt so bleib | mit der langen schneid starckh auff dem schwertt hebt | er den mit dem schwertt vbersich so spring mit deim | lingken fus ~~vor~~ hinder sein Rechten vnd schlag in | mit der zwerch oder sunst zu dem kopff seiner Rechten | Seitten vnd arwait

side. Watch if you can set on to him. But if he strikes at you above from his left side, then shove the point into the opening of his right side.

Item: For he who strikes to you upwards from below from his right side, shove the point into him and set on. Do the same thing when he strikes up from the left side. If he fends off the setting on, then displace and remain with your sword on his. Work nimbly towards the next opening; in this way, you won't let him come to any art.

The text about travelling after

Learn travelling after in two ways or cut into the defence. Take two outer ones; after that, begin your work. Check the encounters whether they are soft or hard.

Note the gloss: There are many different ways of travelling after that have to be performed with great care against those fighters that move about and fence with long strikes or don't care much about the true art of the sword.

Item: Make the first piece from the travelling after like this: When you come with the *Zufechten*, then put the left foot forward and stand in the guard *vom Tag*. Watch carefully for what he fights against you. If he strikes at your left above from the shoulder, then don't displace that, but take care that he doesn't reach you with the strike and press him. Note: While his sword goes down to the ground with the strike, jump towards the opening on his right side before he comes up with his sword. Thus he is beaten.

This piece is called the iron gate

Item: When his strike misses, you strike after him and he goes instantly up with the sword, then remain with the long edge strongly on his sword. If he raises the sword up, jump with your left foot behind his right and strike with the *Zwerch* or anything else to his head on the right side. Work also to his left side with the *Duplieren*

widerum zu seiner lingken | mit dem dupplieren oder sunst mit andern stuckhen | darnach als du empfindst ob er hertt oder waich[13] im | Schwertt ist

Das ist ein ander stuckh

Item wan er sich vor dir verhawt vnd du ym noch | hawest bindestu den auff sein schwertt gegen seiner | lingken seitten schlöcht er den aus der versatzung bald | vnd mit der zwerch dir zu der Rechten seitten so kum | im Indes mit der zwerch vorn vnder sein schwertt | gegen seiner lingken seitten an sein hals oder spring | mit dem lingken fus auf sein Rechte seitten vnd haw | seinem haw auch nach zu seiner Rechten seitten oder
49R treib den schnidt vber seinen arm zu dem kopff

Hie merckh ein gutt nachRaysen

Item wen du gegen im vichst aus dem vnderhewen oder | aus den straichen oder ligst gegen im in der hutt velt aber | er mit dem schwertt auf das dein Etwa du aufkomest | So bleib vnden am schwertt vnd heb vbersich wil er dir | den aus versatzung einhawen oder am schwertt einwinden | so las in vom schwertt nit abkomen vnd volg im daran nach | vnd arwaitt damit zu seiner blöss am negsten Item merckh | du solt aus allen hutten vnd mit allen hutten nachraisen | also bald du ~~nach~~ komest wen er sich vor dir verhawt | oder emplöst mit dem schwertt

Das ist der text von dem Empfinden vnd | von dem wort das da haist Indes

Das Empfinden lere Indes das wortt schneidet sere gloss | merckh das im schwertt das fullen vnd das wort Indes die | grösten kunst ist vnd wen ain maister ist vnd sein wil vnd | kan er das Empfinden nit fullen vnd vernimpt dartzu | nit des wortt Indes so ist er nit ein maister sun-

13 JWA1, fol. 48r: 'schwach'.

or any other piece, depending on your sensing whether he is hard or soft in the sword.

Another piece

When he misses before you, you strike after him and bind on his sword against his left side so that he strikes from the displacement instantly and with the *Zwerch* to your right side, then come *Indes* with the *Zwerch* in front under his sword against his left side on his neck. Or else jump with the left foot to his right side and strike after his strike also to his right side, or make a cut over his arm towards the head.

Note a good travelling after

Item: When you fence against him with lower strikes or with striking up or you stand against him in the guard and he falls with his sword onto yours before you were able to come up, then remain below at his sword and lift up. If he strives to strike on you from the displacement or wind at the sword, then don't let him come off the sword, but follow him on it and with this work to his closest opening. Item: Note that you shall travel after from all guards and with all guards as soon as you come when he misses before you or exposes himself with the sword.

The text about the sensing and the word which is called *Indes*

Learn the sensing. The word *Indes* cuts severely. Note the gloss. Note that in the sword, the feeling and the word *Indes* are the greatest art. He who is or wants to be a master but cannot feel the sensing and also does not perceive the word *Indes*, then he is not a master but only a buffalo of the sword. Therefore, before everything else you shall learn to do these two things well.

der er | ist ein büffel des schwerts darum so solttu vor allen | Sachen die zway ding wol lerne machen

Item das empfinden merckh also wen du mit dem zu | vechten[14] kumst vnd einer den andern an das schwertt bindt | so soltu inn dem als die schwertt zu samen glützen zu | hand villen ob er waich oder hörtt sey vnd dir angebund*en* | hatt vnd als bald du das empfindst so gedenckh an das | wortt Indes das ist das du in dem selbigen Empfindst | behendigklich solttu arwaitten mit dem schwertt so wirt | er gschlagen Ee Er sein gwar wirtt

Item hie solttu merckhen das das Empfinden vnd das wort | Indes eins on das ander nit mag sein vnd das vernim | also wan du im an sein schwertt bindest so mustu mit | dem wort Indes villen waich oder hertt vnd wan | du hast gfult so mustü aber mit Indes arbaiten also sind | Sy alweg beieinander wen das wortt Indes ist in
49v Allen stucken das vernim also Indes lauft durch aus | dem dupplieren vnd Indes muttiertt Indes wech- | seltt durch Indes lauft durch Indes nimpt das | Schwertt Indes thutt was dein hertz begertt Indes | ist ein scharpff wortt damit all maist*er* des schwerts | verschnidten werden Rott

Das ist der text

Nachraisen zwiuach trift man den altten schnidt mit | macht gloss merckh ist das du die nachraisen solt treib*en* | zu baiden Seitten vnd schneid auch darein bringen vnd | das vernym also wen er sich vor dir verhawtt es sey vor | der lingken oder Rechten seitten so haw im frö lich nach | zu der blöss fertt er den auch auff vnd bindett dit vnden | an das schwertt so merckh als bald ein schwertt an das | ander glutzt so schneid im nach dem hals oder val | vnd nim den schnidt

14 JWA1, fol. 48v: 'zu im kom[st]'.

Item: Note the sensing in this way: When you come [to him] with the *Zufechten* and one binds at the other's sword, then you shall in the moment the swords clash together instantly feel whether he has bound in a soft or hard way. As soon as you sense this, remember the word *Indes* which means that in the same instant, you nimbly sense what you must work with the sword. That's how he is beaten before he even notices it.

Item: Here you shall consider that the sensing and the word *Indes*, the decisive moment, are not distinguishable from each other. Understand this as follows. When you bind at his sword, you have to feel the weakness or strength with the word *Indes*. Once you have felt this, you have to work again with the *Indes*. So they are always together because the word *Indes* is in each piece. Understand this as follows. *Indes* runs through from the *Duplieren*, *Indes* makes *Mutieren*, *Indes* changes through, *Indes* runs through, *Indes* takes the sword, *Indes* does what your heart desires, *Indes* is a sharp word which cuts all the masters of the sword red.

The text

Travel after twice, if one hits the old cut with might. Note the gloss: you shall travel after from both sides and also use cuts with them. Understand this as follows. When he misses before you with a strike, be it before the left or the right side, then strike unhampered again towards the opening. If he also goes up and binds on your sword below, then pay attention to when one sword clashes at another. Then cut to his neck or fall and take the cut.

Das ist der text von dem vberlauffen

Wer vnden Remett vnd vberlauft der wirt oben gschemt | wan es glitzt oben so sterckh dich das beger ich zeloben dein | arwait mach waich oder hertt vnd druckh das zwiuach | glos merckh das ist wen dir ainer vnden zu vicht wie | du in solt vberlauffen das vernim also wen du mit dem | zu vechten kumst hautt er dir den vnden zu das versetz | Im nyt sunder merckh wen sein vnderhaw gegen dir | gatt so haw im von deiner Rechten achseln oben lang ein | vnd scheub im den ort ~~ein~~ zu dem gsicht oder brust vnd | setz im an so mag er dich vnden nitt erlangen fertt | er den von vnden auff vnd[15] versetzt so bleib mit der lan*gen* | Schneid starckh auff dem schwertt vnd arwait behend | igklich zu der negsten blöss

Item vermerckh wen du im starckh auff sein schwertt | bunden hast schlecht er den aus der versatzung vmb | dir zu der andern Seitten so bind im oben mit der
50R Langen schneid starckh auff die schwech seines schwerts | oben ein zu dem kopff vnd arwait im zu der blöss | also widerum treib zu baiden seitten

Das ist der text wie du stich vnd hew | absetzen soltt

Lere absetzen stich kunstlichen letzen wer auf dich | Sticht den ortt drift vnd seiner bricht von baiden | Seitten driff alle mal wilttu schreitten glos merckh | die absetzen dreib also wen du mit dem zuvechten kumst | stelt er sich den gegen dir als er welt stechen so setz den ling*ken* | vus vir vnd stee gegen im im der hutt des pflugs zu | deiner Rechten seitten vnd gib dich blos mit dem lingke | sticht er dir den zu der blös so wind mit dem schwertt | auff dein lingke seitten gegen seinem stich die kurtz schneid | an sein schwertt vnd setz du damit ab das dein ortt alweg | bleib vnd gegen im sten vnd schreitt damit zu mit dem | Rechten fuss vnd stich im zu dem gsicht oder brust

15 'vnd' added above the line.

The text about running over

Who aims below and runs over will be put to shame above. When it clashes above, then strengthen, this I wish to praise. Make your work softly or hard and press it twice. Note the gloss: this explains how to run over someone who fences against you from below. Understand this as follows. When you come to him with the *Zufechten* and he strikes to you below, then do not displace that. Instead, pay attention to when his lower strike goes against you. Then strike from your right shoulder above long to him and shove the point into his face or breast and set it on. That way, he can't reach you below. If he goes up from below and displaces, remain with the long edge strongly on the sword and work nimbly to the next opening.

Item: Note that when you have bound strongly on his sword and he strikes around out of the displacement to your other side, then you shall bind above with the long edge strongly on the weak of his sword above to his head. Work to his opening, and make this to both sides too.

The text about how you shall set off thrusts and strikes

Learn to set off and to artfully impair thrusts. Hit [with] the point he who thrusts to you, and his [point] breaks. Hit each time on both sides when you want to step. Note the gloss. Set off in this way: when you come to him with the *Zufechten* and he positions himself as if he wanted to thrust, then put the left foot forward. Stand against him in the guard of the *Pflug* on your right side and expose an opening on your left. If he thrusts to that opening, wind with the sword to your left side against his thrust, the short edge on his sword. Set off with this so that your point remains always directed against him. At the same time, step towards him with the right foot and thrust to his face or breast.

Item ein ander stuckh wen du stast zu deiner Rechten Seitten | in dem pflug hawt er dir den zu deiner lingken seitten | oben ein zu der blöss so var auff mit dem schwertt | vnd wind damit auff der lingke seitten gegen seinem | ghultz vor dem in dem ochsen vnd drit hinzu mit dem | Rechten fus vnd stich im zu dem gsicht oder brust vnd | die stickh treib von der lingken Seitten auch also aus | dem pflug als von der Rechten seitten

Das ist der text wie du durchwechseln solt

Durchwechsel ler von baiden seitten stich nit ser | wer auff dich binder [!] durchwechsel in schier | vindett glos merckh der durchwechsel ist vyl | vnd mangerlay die magstu treiben aus allenn | hewen gegen den vechtern die darwider versetzen

50v vnd die do hawen zu dem[16] vnd nitt zu den blösen des | leibs die soltu gar wol lernen treiben mit virsich- | ~~s~~ tigkait das man dir icht ansetz oder sunst ein kum | die weil du durchwechselst Item das durchwechseln | treib also wen du mit dem zu vechten komst so haw | im oben langk ein zu dem kopff hawt er den wid*er* | gegen dir zu dem schwertt vnd nit zu dem leib so las | dein ortt mit dem haw vnden durch wischen Ee das | er dir an das schwertt bindt vnd stich Im zu der | andern seitten wirt er des stichs gewar vnd vertt mit | dem schwertt dem stich bald nach mit versatzung | so wechsel aber durch zu der andern seitten vnd treibs | alweg wan er dir mit versatzung nach dem schwert | vertt vnd das treib zu baiden seitten

Item ein anders wen du mit dem zuvechten kumst so | setz den lingken fus vor vnd haltt im den langen ortt | gegen dem gsicht hawtt er dir den von oben oder von | vnden zu dem schwertt vnd wil dir das hinweg | schlagen oder starckh einbinden[17] so lass den ortt vndersich | Syncken[18] vnd stich im zu der andern Seitten vnd das | treib

16 'swert' is missing here as well as in JWA1, fol. 48v.

17 JWA1, fol. 49r: 'anbinden'.

18 The initial 'S' superscribes 'h'.

Item: Another piece. When you stand in the *Pflug* on your right side and he strikes to your left side above to the opening, then go up with your sword and wind it to the left side against his hilt forward into the *Ochs*. Step ahead with the right foot and thrust to his face or breast. Make these pieces from the left side in the same way as from the *Pflug* on the right side.

The text about how to change through

Learn to change through from both sides. Don't thrust too much. He who binds on you will be found open by changing through. Note the gloss: there are many and diverse ways of changing through and you can make them out of all strikes against those fencers who displace against them and who strike to the [sword] and not to the openings of the body. You shall learn to perform them with caution so that nobody sets on you or breaks in while you change through. Item: Make the changing through like this: when you come with the *Zufechten*, then strike above long towards his head. If he strikes against you but to the sword and not to the body, then let your point pass through with the strike before he binds on your sword and thrust him on the other side. If he notices the thrust and immediately follows the thrust with the sword with displacing, then change through another time to the other side. Do this always when he goes to your sword with displacing; and make it on both sides.

Item: Another one. When you come with the *Zufechten*, then put the left foot forward and hold the long point towards his face. If he strikes to your sword either from above or below in order to strike it away or to bind to you strongly, then let the point sink down and thrust to his other side. Do this against all strikes that are directed to your sword and not to the openings of the body.

gegen allen hewen so man dir zu dem schwertt | hawtt vnd nit zu den blösen des leibs Item das stuckh | merckh gar eben wen er dir vor versetzt vnd lass | den ortt laffen schon

Item hie hatt der text ein end von dem zetel des | langen schwerts wie man sich halten sol im | schwertt alle dritt vnd mas vnd stich vnd hew | letzen straich mit stossen vnd auch alle blossen | vnd wan ainer waich ist so bis du starckh vnd | wen ainer starckh ist so bis du waich so vind | estu schwech vnd sterckh mit ein ander gutt inn | der hutt

51R Item es ist zewissen das die neche die kunst bricht | vnd die zway henngen vnnd das verschieben vnd | die hol versatzung vnd die guldin kunst die | funf stuckh die brechen den zettel auch vindt | man wol ain der in bricht mit aim oder | zwayen stuckhen den es bricht ein haw den | Anndern vnd ein stuckh das Annder vnd | ein stich bricht den andern gloss merckh

I R

Hie endett sich Jörg wilhalms | hutters kunst zu augspurg etc | des langen[19] schwertt
1523

19 'langen' is absent in JWA1, fol. 49r.

Item: Pay close attention to this piece when he displaces you in front, then let your point run thoroughly.

Item: Here ends the text of the Notes of the long sword about how to hold yourself with the sword. All steps and measurements, thrusts and strikes cause harm to strikes with hits and also all openings. When someone is soft, then you are strong; and when someone is strong, then you are soft. This is how you will find both weakness and strength properly in the guard.

Item: It is important to know that proximity breaks the art. The two hangings and the shoving away, the hollow displacement, and the golden art: these five pieces break the Notes. One may also find somebody who breaks it with one or two pieces because one strike breaks another, and one piece another, and one thrust another. Note the gloss.

Here ends the art of the long sword by Jörg Wilhalm, hatter in Augsburg.
1523

Jörg Wilhalm

Armoured Combat

59R I 1522 P
Hie stand drey p*er*son ain Reich*er* ain starcker ain schwacher
[1]

Wan hatt der Reich | weder kunst noch | witz so vertut er | hab vnd gutt vnd | ist im gar nichts | nitz

Wan hat der starck | kunst noch mutt | So vberkumpt er | weder hab noch gut | war zu ist im sein | Sterckh gutt

Wen hatt der schwach | kunst vnd witz die mag | er brauchen mit gutt*en* | Sinen damit mag er | gross er vnd gutt | gewinen

59V Hilff herr du ewigs | wortt hilff hie vnd | der Sel dortt

60R [2]Hi~~lff~~ie fiertt der maister Sein heren Ein In die Schranckn
wie er Sich halltten Sol

60V Hie komend die zwen in die schrancken zu Samenn | vnd wie Sy den kampff volenden wöllen

61R **Das ist die erst hut vnd Stand**

Item So du ainen zu kampff lernen wilt So merck wie | er gesipt Sey oder ein gut hertz hab das er in vertrau zu | fechten vnd ob er gerecht Sey oder nit ist er gerecht So weiss | in in die Recht kunst leg in an ain schweren harnasch an[3] | damit er brauchsann werd zum ernst sol er Ringen harnasch | haben das er in brauchen mag wirt ain kampff gehabt oder | gesezt zu den dreien weren mit dem spiess oder schwertt | vnd dolch So lern in ~~de~~ zum ersten mallen die zwen

1 Note in pencil from a later hand – quite likely a librarian's: 'cf. Cgm 3712 f 156'.

2 JWM2 only starts here; the earlier sections of the work are absent.

3 JWM2, fol. 2r: 'dar'.

I 1522 P

Three persons are standing here: a rich one, a strong one and a weak one.

If a rich one possesses neither art nor wit, he wastes goods and chattels, and there is nothing that avails him.

When the strong one possesses neither art nor courage, he can acquire neither goods nor chattels. So what is his strength good for, anyway?

When the weak one possesses art and wit, he may use it well with clever senses. In this way, he can win greatly and worthily.

Help, Lord, you eternal Word; help here and help the soul there.

The master leads his lord into the lists. This is how he shall conduct himself.

The two come together in the lists where they want to accomplish the fight.

The first guard and stance

Item: If you want to teach someone how to fight, then observe his attitude. If he has a valiant heart, then trust him to fight. Watch whether he is a just man or not. If he is just, then instruct him in the rightful art. Dress him in a heavy harness so that he becomes used to it for the combat. If he is about to wrestle, he shall have a suitable harness. If a duel is scheduled or announced for the three weapons – spear, sword and dagger – then teach him to begin with the two stances and guards as they are painted above so that he will be able to assume them reliably.

61v stend | vnd hutt wie oben gmalt statt das er in stand gwiss Sey

Das ander Stuck

Item die ander zwen stend oder hutt wie vor gschriben stat | merck gar eben das du in deiner arwait gwiss seiest | vnd wolbedacht wie du in den man wellest arwaitten | vnd senck dich wol gen im wie oben gmalt statt

62r

Das dritt Stuck

Item vermerck den dritten stand der haist[4] die sechst hutt so | einer den knopff gegen dir kertt vnd hatt das schwertt auff | der lingkhen Seitten So Senck dich wol mit deinem | ortt gegen der Erdtt gen im wie oben gmaltt stat schau | das du dich nit verfieren lassest dan es gatt gar hubsche | arwait darauss

62v

Das viertt haist die haftung

Item vermerck So ainer auff dich arwait als vorgeschriben | stat So schaw du gar eben das du mechst kumen auff die | hafftung oder ansetzen auff seinem fuss welcher im vor stet | drifst du in so fal mit deiner prust auff das schwertt vnd | druck es wol nider So stichst du ~~vss~~ in[5] wol nider durch den | fuss wie oben gmalt ist das ist das erst ansetzen oder arwait | in dem kampff

63r

Das vunft haist anwerffen

Item So du sichst das dir ainer wil ansetzen nach dem fuss | wie vor gmaltt Statt So ge auff mit deinem Schwertt | Stoss im nach dem gsicht vnd dritt mit dem andern fuss | vir So prichst du im die haftung wie oben gmaltt statt | vnd Setz dich wol

4 In JWM2, the beginning reads: 'Item vermerck das drit Stuck der haist ...'

5 'in' is inserted above the deletion.

The second piece

Item: The second two stances or guards as written above. Pay close attention that you are sure in your work and that you have reasoned well how to work against the adversary. Sink properly down when facing him as shown in the painting above.

The third piece

Item: Note the third stance which is called the sixth guard. When someone turns his pommel to you, holding the sword on the left side, then sink down, with your point to the ground, as painted above. Watch out that you don't let him deceive you, for there is much skilful work that can be made from this.

The fourth is called the pinning

Item: Pay attention when someone works against you as written above, then watch carefully that you can achieve the pinning or setting on on his foot which he has put forward. If you hit, fall with your chest on the sword and press it down properly. This is how you thrust well down through his foot as can be seen in the painting above. This is the first setting on or the work in combat.

The fifth is called throwing on

Item: If you see that someone wants to set on your foot as painted before, then go up with your sword. Thrust to his face and step forward with the other foot. This is how you break his pinning as can be seen painted above. Position yourself suitably.

63v

Das Sechst ist die haftung

Item So~~n~~ ainer vmb den obern stich wie oben gmalt ist | nichsth ge-ben wil vnd sticht dir nach dem hindern fuss so | fal mit deinem schwertt[6] mit dein baidem henden auff | Sein schwert[7] trucks im ni-der So machst du dich ledig | wie hernoch gschriben statt vnd bis ge-wiß in der arwait | wie oben gmalt Statt

64r

Das Sibent haist der mordschlag

Item die drit haftung oder ansetzung[8] So du das gwar | wirst So nim das schwertt bey dem ortt vnd ker der[9] knopff | vir schlag im zu Seim helm So wirtt er temisch in seim | helm wie oben gmalt Stett So wirst du ledig

64v

Das acht haist haftung

Item So dir ainer nach dem kopff schlechtt oder schlagen wil | So gang auff mit deinem schwertt vnd setz im an mit | dem ortt an dem[10] tengken elenpogen damit hast du im | Sein mordschlag brochen vnd ge wol in dich vnd scheub | in[11] die höch wie oben gmalt Statt

65r

Das neundt vberwinden

Item vermerck die neundt arwaytt So dir ainer an | gesetz hast vnd wil dich schieben wol zu Rugk So nim | dein Schwert bei dem ghiltz in die tengken handt vnd | drit mit deinem Rechten fuss vir vnd wind im mit | dem knopff nach den gsichtt So prichst du im das an | Setzen vnd haftung wie vor gmalt Statt vnd die | arwait der bruch die darauff stett[12]

---

6 Unclear reading due to an ink stain, possibly corrected from 'schwerth'.
7 See previous footnote.
8 JWM2, fol. 5r: 'nach dem fuß'.
9 JWM2, fol. 5r: 'den'.
10 JWM2, fol. 5r: 'Sein'.
11 JWM2, fol. 5r: 'in in'.
12 JWM2, fol. 6r: '... wie gmalt ist'.

The sixth is the pinning

Item: If someone pays no attention to an upper thrust as it is painted above and he thrusts to your rear foot, then fall with your sword with both hands on his sword and press it down. In this way, you liberate yourself as it is written hereafter. Be confident in your work as it is painted above.

The seventh is called the murder strike

Item: The third pinning or setting on [to the foot]. If you notice this, take your sword at the point and turn the pommel forward. Strike to his helmet, so he'll be dazed in his helmet, as painted above. That's how you liberate yourself.

The eighth is called pinning

Item: If someone strikes to your head or tries to strike there, go up with your sword and set on his left elbow with the point. With this, you have broken his murder strike. Brace yourself properly and lift up high, as it is painted above.

The ninth is winding over

Item: Pay attention to the ninth work. When someone has set on you and tries to push you thoroughly backwards, then take the sword at the hilt in the left hand and step forward with your right foot. Wind with your pommel to his face; in this way you break his setting on and pinning as it has been painted before. This is the work and the break against it.

65v **Das zechent Stuck**

Item So du Sichst das dir ainer nach dem gsicht arwait | So drit vir mit deinem rechten fuss gee mit deinem | Schwertt wol in die höch vnd stoss mit deim schwertt | inerhalb Seins schwertts vber Sein Rechten arm das haist | zwischen baiden armen durch gestochen vnd druck wol | nider drit mit deim tengken fuss vir So hast du das | ansetzen oder haftung auff Seinem Rechten fuss[13]

66r **Das aylft Stuck**

Item So du sichst das dir ainer zwischen baiden armen | ~~w~~ durch wil stossen vnd dir dis sterck hat abgwun*n*en | So drit vir mit deim teng-ken fuss vnd wol auff mit | deim ortt nach seinem gsicht wie oben gmalt Stett So | hast du in geprochen

66v **Das zwelft Stuck**

Item So du sichst das dir ainer nach dem gsicht arwait | So drit vir mit deim Rechten fuss vnd wind mit deinem | knopff mit deiner Rechten hand vnder Seiner tengken | mit dem schwertt zeuch wol an dich wie du oben vor dir | gmaltt sichst

67r **Das dreitzechenst Stuck**

Item So du sichst das dir ainer durch windt wie vor | gmalt vnd gschri-ben Statt So drit mit deinem tengk*en* | fuss fir vnd begreiff Sein schwertt mit samptt deine[14] | in dein lingke hand wind wol nider das haist der | klemer wie oben gmalt Statt

13 JWM2, fol. 6v: '… wie oben gmalt Stett'.

14 JWM2, fol. 8r: 'dem deinen'.

The tenth piece
Item: When you see that someone works to your face, then step forward with your right foot. Go high up with your sword and beat with your sword on the inside of his sword over his right arm. This means thrusting through between his arms. Press properly down. Step forward with your left foot, so you have the setting on or the pinning on his right foot.

The eleventh piece
Item: When you see that someone wants to thrust through your arms, having taken your strength, then step forward with your left foot and go up with your point to his face, as in the painting above. In this way, you have broken him.

The twelfth piece
Item: When you see that someone works to your face, then step forward with your right foot and wind with your pommel with your right hand under his left. Pull with the sword decisively towards you, as you see in the painting above.

The thirteenth piece
Item: When you see that someone winds through you as it is painted and written before, then step forward with your left foot and grasp his sword together with yours in your left hand, wind properly down. This is called the clamper, as painted above.

67v **Das vierzechenst Stuck**

Item[15] So du Sichst das dir ainer wil fallen in den klemer | So drit mit deinem Rechten fuss vir wind mit deim schw | ertt vnd knopff wol auff das du im an den hals kumest | zeuch wol zu Rugk So wirffstu in vber den fuss wie | oben gmalt Stett

68r **Das vunffzechenst Stuck**

Item so dir ainer nach der vnder haftung oder ansetzen oder | wil dich ~~werffen~~ vber die pain werffen So thu im also | gang auff mit deim schwertt vnd drit wol vir hinder | Sein fuss vnd fall im mit deim schwertt auff Sein schwert | vnd truck Ims wol nider auff den hals wie oben gmalt | Statt das ist der viertt hauptt prich ainer

68v **Das Sechzechenst Stuck**

Item So dir ainer vmb den halss wnnt [!] vnd wil dich werff*en* | So zuck den fuss an dich vnd dritt wider vir mit deinem | Rechten hinder Seinen Rechten vnd wind mit deinem | lingk*en* wol nider dein lingken elenpogen hinder Sein Rech*en* [!] | Ichssen druck von dir So wirfstu in auff das angsichtt | wie oben gmaltt statt das ist der ander hauptbruch

69r **Das Sibenzechenst Stuck**

Item So du sichst vnd empfindt das dich ainer also vertruck | enn wil So drit mit deinem Rechten fuss fur ausser halb | Sein Rechten nim dein tengke hand vnd greiff an Sein | Rechten elenpogen Stoss wol von dir So hast du im das | vorder Stuck gebrochen[16] wie oben gmalt Statt

15 JWM2, fol. 8v: 'Item das vierzechnest Stuck'.
16 JWM2, fol. 10r: 'das Sichst du wol'.

The fourteenth piece

Item: When you see that someone wants to perform the clamper, then step forward with your right foot. Wind well up with your sword and pommel, so that you come to his neck and pull adequately backwards. This is how you throw him over the foot as it is painted above.

The fifteenth piece

Item: If someone aims for the lower pinning or setting on or if he tries to throw you over the leg, then do the same to him. Go up with your sword and step decisively behind his foot and fall with your sword on him. Press it well down on the neck, as it is painted above. This is one of the four principal breaks.

The sixteenth piece

Item: When someone winds around your neck and strives to throw you, then twitch the foot towards you and step again forward with your right behind his right. Wind with your left properly down with your left elbow behind his right armpit. Press away from you; this is how you throw him on his face, as it is painted above. This is the second principal break.

The seventeenth piece

Item: When you see and sense that someone wants to press you in this way, then step with your right foot forward on the outside of his right. With your left hand, grasp his right elbow. Push decisively away from you, and so you have broken the previous piece, as [you can well see] in the painting above.

69v **Das achtzechenst Stuck**

Item So dich ainer hinweg stost mit dem lingken dein*n* | Rechten el-lenpogen So gee auff mit deinem Rechten arm | vnd mit deim teng-ken Sein tengke bey dem ellenpogen | mit deiner Rechten handt vmb Sein halss nim in wol | bey Seinem partt oder visier zem in wol in die höch | oder puck dich nider welches du wiltt das ist vast ain | gutter pruch vnd Stuck wie oben gmaltt Statt So | magst du die ler wol be-haltten

70r **Das nonzechenst Stuck**

Item So du Sihst das dir ainer vmb den halss wil fallen | vnd wil dich mit Seiner Rechten dein tengken ellenpogen | weck stossen So dritt mit deinem Rechten fuss fir hinder[17] | Sein knie mit deiner lingke handt innen[18] zwischen Seiner | pain in der kniepug heb wol in die höch ~~R~~ sein Rechten | fuss So wirfstu in an den Rugken wie oben gmaltt | Statt

70v **Das zwaintzigst Stuck**

Item So du sichst das dich ainer also werffen wil wie oben | gmalt Stet So greiff mit deiner Rechten oben vber Sein | halss in Sein Rechten ellenpogen zeuch dein tengken arm | wol an dich vnd leger dich wol auff in So prichst du | im das voder Stuck vnd druck in nider zu der erdt | wie oben gmaltt Statt das ist vast ain gutter | pruch

71r **Das ainundzwaintzigist Stuck**

Item So du Sichst oder empfindst das er dich zu der erd | trucken wil So greiff mit deiner Rechten in Sein Rechts | glenck vornen bey der hand vnd heb mit deiner Rechte | Sein Rechte hand auff vber Sein[19] halss vnd drit mit deinem | tengken fuss vir vnd greiff mit deiner

17 JWM2, fol. 11r: 'Sein Recht*en* vnd greiff mit deiner Rechten handt aussen an'.

18 Actually, there is one stroke too many to correctly read 'innen'.

19 JWM2, fol. 12r: 'den'.

The eighteenth piece

Item: When someone pushes your right elbow away with the left, then go up with your right arm and [grasp] with your left his left at the elbow and [go] with the right hand around his neck. Grab his bevor or visor firmly, pull him up high or bend down, whatever you wish. This is a very strong break and piece, as it is painted above. Remember this lesson well.

The nineteenth piece

Item: When you see that someone wants to fall around your neck and wants to push away your left elbow with his right hand, then step with your right foot forward behind his [right and grasp with your right hand to the outside of his] knee, [go] with your left hand between his legs to the inside into the knee bend. Lift his right foot well up, in this way you will throw him on the back, as in the painting above.

The twentieth piece

Item: When you see that someone wants to throw you in the way as it is painted above, then grasp with your right above over his neck into his right elbow. Pull your left arm properly towards you and place yourself decisively on him. In this way, you break his previous piece. Press him down to the ground as painted above. This is a very effective break.

The twenty-first piece

Item: When you see or sense that he wants to press you to the ground, then grasp with your right in his right wrist and lift up with your right his right hand over his neck. Step with your left foot forward and grasp with your left hand in his harness on the back, in his buttocks or between his legs, whatever may happen.

tengken hand hinden | in Sein harnasch oder gesess oder zwischen der pain welches | dir werden mag höb wol in die höch So wirfst du in | an den Rugken wie oben gmalt Statt vnd ist vast | ain gut Stuck

71V **Das zweiundundzwaintzigist Stuck**

Item So er dich an den Rugken werffen wil So greiff | mit deinen baiden armen oben Seiner achsel vnd greiff | mit der ~~Rechten~~ tengken[20] in Sein visier vnd greiff du mit | deiner Rechten in Sein tengke vnd So er dich gfast hatt | mit der lingke hand bey deinem tengke pain vnd | wil dich heben zum wurff So leger dich wol auff in | truck in zu der erd wie oben gmaltt Statt So hastu | im sein arwait prachen

72R **Das dreiundzwaintzigst Stuck**

Item So dich ainer also nider trucktt zu der Erdt wie vor | gmalt[21] Stat so fal fluchs hindersich auff den Rugken | So muss er mit dir fallen So wirst du ledig von Im | lig nit lang spring bald auff vnd stoss mit deim Rechte | knie an Sein tengke fuss So muss er von dir lassen wie | oben gmaltt Statt

72V **Das vierundzwaintzigst Stuck**

Item So du nun auffgesprungen büst eil bald zu deinem | Schwertt ober sich versaumpt So hast du gut arwaiten | ist er behend als du vnd erwist Sein Schwertt auch so | eil mit deinem langen Stich auff in vnd begreiff mit | deiner tengken handt Sein ortt begreiftt er das dein auch | bey dem ortt So stee fest schau ob er starck oder schwach | Sey darnach waist du zu arwaiten wie oben gmalt | Statt das haist der anfang der scherr

20 In JWM2, fol. 12v, the cancellation 'Rechten' remains and has not been replaced by 'tengken'.

21 JWM2, fol. 13r: 'vnd geschriben'.

Lift him well up high, this is how to throw him on the back, as painted above. This is a very efficient piece.

The twenty-second piece

Item: When he strives to throw you on your back, then grasp with both arms over his shoulder with the left in his visor and with your right in his left. If he has grasped you with the left hand at your left leg and strives to lift you in order to throw you, then place yourself well on him and press him down to the ground, as painted above. This is how you have broken his work.

The twenty-third piece

Item: When someone presses you down to the ground as painted [and written] above, then fall quickly behind you on your back. In this way, he must fall down together with you. This is how you liberate yourself from him. Don't lie there for a long time but jump up instantly. Hit his left foot with your right knee so he has to let you go, as in the painting above.

The twenty-fourth piece

Item: When you have jumped up, hurry instantly to your sword when he doesn't react. In this way you can work efficiently. If he is quicker than you and also gets his sword, then hasten with a long thrust to him and grasp his point with your left hand. If he grabs your point too, stand firmly and observe whether he is strong or weak. You know how to work accordingly, as in the painting above. This is called the beginning of the scissors.

73R **Das vunffundzwaintzigist Stuck**

Item So du empfindst das baid Schwertt gfangen Sein wie | vor stett So wind fluchs durch mit dein baid armen kreutz | weiss vber ain ander nimss auff dein achssel buck dich | nider vnd zeuch wol an dich baide schwertt wie oben | gmalt Statt vnd kerr im den Rugken das haist die | Schranckhutt

73V **Das Sechsundzwaintzigist Stuck**

Item So ainer also auffgwunden hatt in die Scher vnd wil | dir dein Schwertt nemen So las im baiden schwertt | mit willen vnd faln mit deim kopff in Sein Rugk*en* | vnd greiff mit baiden henden vnden in Sein fuess an | die kniepug zeuch wol an dich vnd mit dem kopff | Stoss von dir So wirfstu in auff den pauch wie oben | gmaltt Stett

74R **Das Sibenundtzwaintzigist Stuck**

Item So dich ainer werffen wil wie vor stet So drit mit | deim tengken fuss fir mit deim Rechten ker dich bald vmb | vnd stoss in mit dem knopff zu Seim helm So du das | gwar wirst vnd get dir auss der arwait vnd wil dir nach | dem helm stossen mit dem knopff So ge auff mit deiner | tengken hand ann sein Rechte vornen an das ghultz So | brichst du im den stos greiff bald mit deiner Rechten hand | nach deinem schwertt So kumpst du wider in die arwait

74V **Der schirenprandt**

Das achtundzwaintzigist Stuck

Item So ir gegen ain ander kumpt in die arwait So ste | vest vnd schau ob er starck auff dich wel lauffen oder nit | So ge mit deinem schwertt vor im vmb im wirbel vnd | thu als wellest du in verfieren das schtuck haist der | Schirenbrand vnd gat vil arwait drauss

The twenty-fifth piece

Item: When you sense that both swords are caught as is written before, then turn quickly through with both arms, crosswise one over the other. Take it on your shoulder, bend down and pull both swords properly towards you, as it is painted above. Turn the back towards him. This is called the *Schranckhut*.

The twenty-sixth piece

Item: When one has turned up in this way into the scissors and tries to take your sword, then deliberately let him have both swords. Fall with your head in his back and grasp with both hands below his feet in the knee-bend. Pull well towards you and push him with your head away from you. This is how you throw him on the belly, as in the painting above.

The twenty-seventh piece

Item: When someone tries to throw you as written before, then step with your left foot forward. Immediately turn around with your right. Hit to his helmet with the pommel. When you notice this and he avoids your work in order to hit to your helmet with the pommel, then go up with your left hand to his right, in front at the hilt. In this way, you break his strike. Reach instantly with your right hand to your sword, so you can return to your work.

The sword chafing

The twenty-eighth piece

Item: When you encounter each other in your work, then stand firmly and observe whether he wants to run strongly to you or not. Then move your sword in front of him around in a twist and pretend to deceive him. This piece is called the sword chafing, and much work can be made with it.

75R **Das neunundzwaintzigist Stuck**

Item So ainer also Starck in dich wil laffen vnd stost | dir oben nach dem gesicht vnd bergert der haftung oder | ansetzen vnd stett vor mit Seim tengken fuss[22] vnd // | So dritt | du vir | mit deim | tengken | fuss | // ge auff mit deiner lingke handt an dein[23] schwertt ~~wie~~ | an Sein tengke handt an Seim schwertt wie oben gmalt | Stett vnd merck den dritt gar wol

75V **Das dreissigist Stuck**

Item So ainer Starck auff dich lauf vnd wil dir die vnder | haftung oder ansetzung begertt vnd stet vor mit Seinem | tengke So drit du mit deinem Rechten fuss vir zu Seim | lingken vnd val mit deim schwertt mit deiner lingke | hand an Sein schwertt bei Seiner lingke handtt truck | vnden wol von diyr So stost du in mit dem knopff in das | gsicht wie oben gmaltt Stett ist vast ain gutt Stuck

76R **Das ainundreisigist Stuck**

Item So dich ainer vbereilt vnd kanst zu kainer arwait | kumen So greiff mit deiner lingke handt in sein ghultz | bey dem knopff[24] vnd Rechten hand in das viesier wie oben | gmalt Stett das Stuck haist die ledigung

76V **Das zweiundreisigist Stuck**

Item So ainer dir nach dem gsicht wil stossen So senck | dich nider vnd dritt mit deim Rechten fuss vir vnd | mit deinem knopff vnd haggken greiff in oder gee durch | in sein kniepug heb auff den fuss So wirfst du in an den | Rugken wie oben gmalt Statt das stuck gatt von baiden | Seitten oder arwait von baiden seift [!]

---

22 The passage 'So dritt ... tengken fuss' has been added at the left margin. An according insertion mark in the form of two double slashes // can be found in the text, but in the wrong position: once after 'vnd' and one before 'ge'. See also JWM2, fol. 16r.

23 'd' corrects a previously written 'S'.

24 JWM2, fol. 17r: 'zeuchs im wol nider vnd Stoß mit deinem knopf'.

The twenty-ninth piece

Item: When someone wants to run strongly into you in this way and thrusts above to your face in order to pin you or set on you while having put his left foot forward, then step forward with your left foot. Go up with your left hand with your sword to his left hand on the sword, as painted above, and remember well to make a step.

The thirtieth piece

Item: When someone runs to you strongly and tries to use a lower pinning or setting on while having put forward his left [foot], then step forward with your right foot towards his left and fall with your sword with your left hand at his sword close to his left hand. Press properly down away from you, that's how you hit him with the pommel in the face, as painted above. This is a very efficient piece.

The thirty-first piece

Item: When someone overpowers you and you can't come to your work, then grasp his hilt at the pommel with your left hand [pull it well down and hit with your pommel] and your right in his visor, as it is painted above. This is called a liberation.

The thirty-second piece

Item: When someone strives to thrust to your face, then sink down and step with your right foot forward and grasp him with your pommel and hook. Or else go through in his knee bend and lift up the foot. This is how you throw him on the back, as painted above. This piece can be made on both sides, or work on both sides.

77R
Das dreiunddreisigist Stuck

Item So sich ainer senckt nach dir vnd fertt dir durch | die pain vnd wil dich werffen So zuck dein Rechten | fuss an dich vnd fal im mit deim schwertt ann sein | halss wie oben gmalt stett So hastu im das voder | Stuck geprochen vnd druck wol nider

77V
Das vierunddreisigist Stuck

Item So dir ainer das schwertt also an den hals glegt | hatt So ge vnden auff durch Sein schwertt wind mit | deinem schwertt wol auff dritt mit deinem tengken | fuss vir hinder Seine pain leg im das schwertt an | den hals vnd mit deinem elenpogen stoss Im an sein | go-der Zeuch wol hindersich wie oben gmalt statt so | wirfstu in vber das pain an den Rugken

78R
Das vunffundreissigist Stuck

Item So du Sichst vnd empfindst in der arwait das er | dich werffen wil vnd dir das schwertt an den hals glegt | hatt wie vor statt So drit mit deim Rechten fuss vir | vnd wind mit deim schwertt durch Sein baid arm das der | knopff vnd das ghulz vor stee truck wol nider vnd | Senck dich So prichst du im den wurff wie oben ~~gm~~ | gmaltt Statt das ist die viertt arwaitt aine Im | kampff

78V
Das Sechsunddreisigist Stuck

Item vermerck die Rechten stend vnd arwait So dich | ainer also ni-dertruckt wie vor stett so ste auff mitt | deim schwertt vnd lug[25] wie du mitt den fiessen ledig | mugst werden das kan man also gruntlich nit schreib*en* | oder malen So du auff bist gangen mit dem schwertt | vber den kopff Stoss im flux nach der haftung wo der | helm auff ain ander gatt das haist die haftung oder | ansetzen so stost du in wund wie oben gmalt statt

25 JWM2, fol. 19v: 'merck'.

The thirty-third piece

Item: When someone sinks towards you and goes through your legs in order to throw you, then twitch your right foot towards you and fall with your sword on his neck, as in the painting above. In this way, you have broken the previous piece. Press properly down.

The thirty-fourth piece

Item: When someone has placed his sword on your neck in this way, then go up from below through his sword. Wind with your sword decisively up. Step with your left foot behind his legs and put the sword on his neck. Push him with your elbow at his throat. Pull properly backwards, as painted above. That's how you throw him over the leg on his back.

The thirty-fifth piece

Item: When you observe and sense in your work that he wants to throw you and he has placed his sword on your neck as written above, then step forward with your right foot and wind with your sword through both his arms so that the pommel and the hilt are directed forwards. Press well down and sink down, this is how you break his throw, as it's painted above. This is the fourth work in combat.

The thirty-sixth piece

Item: Notice the correct stances and the work when someone presses you down as written above. Go up with your sword and observe how you can liberate yourself with the feet, which one can neither describe nor paint exhaustively enough. When you've gone up with the sword over the head, push him quickly for a pinning where the helmet is placed. This is called the pinning or setting on. This is how your hit injures him, as painted above.

79R **Das Sibenunddreisigist Stuck**

Item So[26] du sichst das dich ainer also wil stossen nach | der haftung an dem hals So setz dich nider vnd Senck | dich wol nider in die mass vnd leng vnd stoss im | vnder die Rechten yechssen da der harnasch tailt ist | So stost du In durch das pantzer blutriss[27] wie oben | gmaltt ist darnach finst du gutte arwaytt

79V **Das achtundreissigist Stuck**

Item So du also Sichst das dir ainer die haftung vnd | ansetzung sucht vnder deiner Rechten yechsen vnd dein | stoss nit verbringen magst wie vor stett so fall behend | mit deiner gewapententen [!] hand vnd schwertt auff | Sein lingke hand truck wol nider wie oben gmalt | Statt ist vast ayn gutter bruch

80R **Das neununddreissigist Stuck**

Item So dir ainer verfeltt nach der arm wie vor statt | So lass dein schwertt fallen vnd greiff mit deiner | lingke handt auff sein Rechte hand vnd drit vir mit | deinem Rechten fuss vnd greiff mit deiner Rechten | hand zu seinem Rechten pain in die kniepug zuck dein | fuss an dich oder heb in auff vnd druck mit deiner | achssel wol dir So wirfstu in an den Rugken wie | oben gmalt Statt ist vast ain gutt stuck in dem | Ringen zu dem kampff von baiden Seitten gettes

80V **Das viertzigist Stuck**

Item So dir ainer also vnderlaft in dem Ringen | vnd wil dich werffen So mach dich ledig mit den fies*en* | vnd greiff mit deiner Rechten hand vnder Sein yechsen | vnd greiff mit deiner tengken hand vber Sein Rugk*en* | in das gesess So wirfstu in an den Rugken wie oben | gmalt Stett vnd hernach findst in den kunsten | vnd biss gewiss in der arwalt [!] es Sey im schwertt oder | Im Ringen

26 'S' overwrites an initially written 'd'.

27 JWM2, fol. 20r: 'blutrig'.

The thirty-seventh piece
Item: When you see that someone wants to hit you with a pinning at the neck, then go down and sink properly down in the [correct] measure and length. Hit him under the right armpit where the harness is segmented. This is how you hit him bloody, as it's painted above. After that, you'll find more efficient work.

The thirty-eighth piece
Item: When you see that someone looks for a pinning or setting on under your right armpit and you can't execute your hit as before, then fall nimbly with your armoured hand and the sword on his left hand. Press properly down, as painted above, which is a very efficient break.

The thirty-ninth piece
Item: When someone misses the arm as written before, then drop your sword and grasp with your left hand on his right hand. Step forward with your right foot and grasp with your right hand to his right leg in his knee bend. Twitch with your foot towards you or lift it up. Press with your shoulder strongly away from you, that's how you throw him on the back as painted above. This is a very efficient piece for wrestling in combat, and it works on both sides.

The fortieth piece
Item: When someone runs under you in this way with wrestling and strives to throw you, then liberate yourself with the feet. Grasp with your right hand under his armpit and with your left hand over his back in his buttocks. This is how you throw him on the back as in the painting above and as you will find it hereafter in the arts. Be secure in your work, be it with the sword or in wrestling.

81R **Das ainundviertzigest Stuck**

Item So du ainen also auff den Rugken hast gworffen | wie vorstet So schau bald das du auff in kumpst vnd | mit deim dolch die haftung suchest wo du Sy gewinen | magst So du in gewunen~~e~~ hast wie oben gmaltt Statt | So sprich Im zu ob er sich dem haligen grab geben | wel oder nit

81V **Das zwayundviertzigist Stuck**

Item So du empfindst das dich ainer an den Rugken | gworffen hatt so stoss mit deiner Rechten hand oder mit | deiner lingke mit welcher es dir am ersten kumen | mag vnd stoss in an sein Rechten elenpogen oder begreiff | in halt in vest[28] w~~v~~nd wind dich behend vmb so | wirfs-tu in auff den pauch So Sitz behend auff In | vnd schau wo du in gwi-nen magst wie oben | gmalt statt das ist das pest stuck im kampff zu | Retigung seins lebens

82R **Das dreiundviertzigist Stuck**

Item So du also auff aim sitzst vnd er wil sich vnder dir | auffwinden wie vor gschriben vnd gmalt statt So | greiff mit deiner Rechten in sein tengken ellenpogen | vnd mit deiner tengken greiff im auff[29] Sein Rechte | So er an dem Rugken ligt wie oben gmalt Statt | So werstu im dass er vnder dir nit auff mag sten | du must deine pain wol in hutt haben das er dirss | nit begreiff er wurd sunst ledig vor dir das send | Die besten stuck Im kampff

82V **Das vierundfiertzigest Stuck**

Item So er sich also vnder dir die hatt vmgewunden | So heb in vest mit deinen armen auff Sein arm | wirff in vmb auff den pauch knie hinden mit deine*m* | knie auff sein fiess vnd stich mit deinem dolch

28 The scribe of JWM2 repeats here a passage from the same paragraph in almost identical words: 'welches du amm negsten kumen magst.'

29 JWM2, fol. 23r has 'in' instead of 'auff'.

The forty-first piece

Item: When you have thrown someone on the back as described before, then take care that you get instantly on him in order to look for a pinning with your dagger wherever you can obtain one. Once you have reached one as in the painting above, speak to him as to whether or not he desires to proceed to the holy grave.

The forty-second piece

Item: When you sense that someone has thrown you on the back, then punch with your right hand or with your left one, whichever you can in the first place, and hit his right elbow. Or else grab him and hold tight. Turn nimbly around, that's how you throw him on the belly. Sit quickly on him and observe where you can overcome him as painted above. This is the best piece in combat for saving the life.

The forty-third piece

Item: When you sit on someone and he wants to turn upwards underneath you as described above, then grasp with your right in his left elbow and with your left on his right. When he lies on his back as it is painted above, then you fend it off so that he can't get up from under you. You have to take good care of your legs so that he doesn't grasp them. Otherwise, he'll liberate himself from you. These are the best pieces for combat.

The forty-fourth piece

Item: When he has turned around under you like this, then hold him tight with your arms. Throw him around on the belly and kneel behind with your knee on his feet.

die | ploss oder haftung ~~d~~ wie du im am negsten gwin[30] | enn magst wie oben gmalt statt das ist vast ain | gutte arwaitt

83r **Das vunffundviertzigist Stuck**

Item So du ainen also vmb hast gworffen vnd kums[t] | auff in vnd wilt im das leben nit nemen So greiff | nach deim oder Seym schwertt welches dir am negst*en* | ist Stoss imss vber Sein Rugken vnder den armen das | baid arm auff dem schwertt ligentt So mag er nit auff | Ston So magst du dein dolch zucken wie oben gmalt | Statt vnd mast[31] in zu sprechen nach deim gwissen

83v **Das Sechsundviertzigest Stuck**

Item So du gwiss wilt Sein in der kunst vnd in dem | kampff den dich dan der maister gelernt hatt vnd ym | das schwertt durchgstossen hast So schau das du ain | Remen oder strick du oder dein maister beii euch habtt[32] | vnd bind Im die arm zu Samen wie oben gmalt stat | ~~Sas~~ So mag er nit aufsten du magst auff in knien | oder Sitzen

84r **Das Sibenundviertzigest Stuck**

Item So dir got das gluck geben hatt vnd wol gelungen ist | vnd in gepunden hast vnd im das leben nit nemen wild | So lass in ligen ge von im gib im den segen wen er mag | nit aufsten wen er mied wirtt So wirtt er sich wol geben | das ist ein bschluss des kampffs

Item vermerck das pest stuck So du klain vnd schwach bist | So schaw so ain starcker mit dir kempft vnd wil in dich | lauffen vnd Sich Seiner sterck tröst vnd dich vberstossen | wil So setz dich wol in die wag vnd mach dich klain | vnd ker die Seitten vir auff welche er lauft So drit | du auss auff die ander Seytten

30 One strike is missing; only two are written here, although it should be three in order to be read as 'in'.

31 JWM2, fol. 24r: 'magst'.

32 JWM2, fol. 24v: 'dir habst'.

Thrust with your dagger to the opening or [use a] pinning, depending on what you can reach first, as in the painting above. This is a very efficient work.

The forty-fifth piece
Item: When you have thrown someone around in this way and you get on him but you don't want to take his life, then reach for his or your sword, whichever is closer to you. Push it over his back under the arms so that both arms rest on the sword. In this way he can't get up and you can pull out your dagger as in the painting above. You can make him talk according to your desires.

The forty-sixth piece
Item: If you want to be secure in the art and in the combat which the master has taught you, and you have pushed the sword through him, then make sure that either you or your master carry a strap or rope. Tie his arms together as in the painting above. In this way, he can't get up and you can kneel or sit on him.

The forty-seventh piece
Item: If God has given you luck and you have succeeded in tying him up, but you don't want to take his life, let him lie there. Go away from him and give him your blessing, for he can't get up. When he becomes tired, he certainly will surrender. This is the end of the fight.

Item: Pay attention to the best piece: If you are small and weak and it happens that a strong one fights you who intends to run into you, relying on his strength, in order to hit you violently, then position yourself properly in the scales and make yourself small. Turn the side forward to which he's running, then step away to the other side.

Jörg Wilhalm

MOUNTED COMBAT

85r Item So du vermercken wilt vnd wilt die Riterlichen kunst | lerrnen zu Ross kempffen vnd fechten So sol der maister dich | lernen wie du dich halten solst zu veld oder zu holtz kumst du | in den holen weg vnd bsorgst du dich So nim dein schwert in die | Rechten hand vnd halt in vndersich neben deinem Sattel des | gleichen thu in ainem holtz vnd ker dein Eissen firsich So magst | du arwaiten in dem gerust oder sunst wie dir die arwait | begegnet als hernach gmalt Stet vnd gar gruntlich ausweist

Item vermerck das erst arwaitten So ainer mit dem spies auff | dich Rent So setz dich wol gegen Im in die hutt vnd wass | du in der hand hast es sein ain kolben oder schwertt So setz | Im den spiess ab wie du dan hernach vinst das ist die | erst hutt vnd ausweisen wie oben gmalt Statt

85v Item So ainer mit dem spiess auff dich Rendt vnd was | du in deiner hand hast schwertt oder kolben ge fluchs | auff vnd behend mit dei-ner Rechten hand an den | spiess vnd nachend bei dir truck in auff dein Rechten | Seitten das haist die absetzen vnd ist der ander an-fang | in der natturlichen arwait als oben gmaltt statt

86r Item So du du [!] abgesetzt hast wie vorstett So Reitt wol | In in vnd stoss in in das gsicht oder wirff in von dem | gaul wie du hernach ge-malt vnd gschriben vinst vnd biss | gar virsichtig das du wissest zu arweitten[1]

1 JWA2, fol. 27r: 'wie oben gemalt Stett'.

Item: If you wish to understand and learn the chivalric art of fighting and fencing on horseback, then the master shall teach you how to conduct yourself on the field or in the woods. If you come to a sunken road and you become worried, then take your sword in the right hand and hold it downward next to your saddle. Do the same in a forest, and turn your iron forward. In this way, you can work with the lance rest or in any other way the work encounters you, as it is painted hereafter and also most thoroughly explained.

Item: Understand the first work. When someone races to you with the lance, then position yourself securely in the guard. Set off his lance with whatever you happen to hold in your hand, be it a club or a sword. You'll find this hereafter and it is the first guard and a repulse as in the painting above.

Item: When someone races to you with the lance, then go swiftly up with whatever you hold in your hand – sword or club – and go nimbly with your right hand at his lance, close to you. Press it to your right side. This is called setting off and it is the beginning of the elemental work as it is painted above.

Item: If you have set off as described above, then ride well into him and hit him in the face or throw him off the horse as you can find painted and written hereafter. Be quite careful that you know how to work.

86v Item So zwen zu samen Reitent vnd habent baid spiess so | Schau das du den spies gar gwiss habest vnd im Renen leg | den deinen Ros[2] auff den kopff vnd senck das Eisen ain wen | ig gegen der erd vnd so die eissen gleich zu Samen kumen | So wechsel vnden durch vnd druck auff dein Rechten seitt*en* | So Renst du Im vnden an den partt oder wo du wilt | das haist der durchwechsel als oben gmalt Statt

87r Item So ainer auff dich Rentt vnd kanst dein spiess in das | gerist nit bringen magst das du vbereilt wurst so wurff | dein spiess bald auff dein tengken Seyten vnd das eisen | gegen der erd vnd setz im den stich ab wie hernach | gmalt stat vnd bis wol in der hutt

87v Item So du den spies begriffen hast vnd abgesetzt So magst | du mit dem pausch oder mit dem eysen das haist | nach der kunst auss dem ortt gearwait So empfindst | was du arwaiten solst auch[3] Sichst wie oben gmalt | Stett

88r Item so du mit dem pausch abgesetzt hast So wirff den | zam an den hacken den du vor dir solt haben an deiner | gurttel vnd schau das du das Ross gewaltig seiest mit deine*m* | leib vnd ge auff mit dem spiess vber deinen kopff das | dein Rechte hand vor stee vnd die lingk hinden stoss in | mit dem pausch vnden an den goder So stost du in von | dem gaul wie vor stett

88v Item So dir ainer zu eilt vnd wil dich mit dem pausch | von dem gaul stossen wie vor gmalt stett so wirff dein | spiess von dir vnd greiff mit deiner Rechten hand in | Sein spiess von dir vnd greiff mit deiner Rechten | hand in sein spies[4] bey dem pausch truck wol von dir | wie oben gmalt Stett So wirfst du in als bald als er | dich

2 Looks more like 'Rof'. Interestingly, JWA2, fol. 27v also says 'Roff'.

3 JWA2, fol. 28v has 'So' instead of 'auch'.

4 'von dir vnd greiff mit deiner Rechten hand in sein spies' erroneously written twice.

Item: When two ride against each other and have their lances, then make sure to hold your lance securely. While racing, place it on your horse's head and lower the iron a bit towards the ground. As soon as the irons meet, change through below and press to your right side. In this way, you dash at his bevor or wherever you want. This is called changing through, as it is painted above.

Item: When someone races at you and you can't get your lance in the lance rest so that you are being overrun, then throw your lance immediately to your left side, the iron towards the ground. Displace his thrust as it is painted hereafter, and be safe in your guard.

Item: When you have grasped the lance and set off, then you can use either the graper or the iron. This is called the work according to the art of the point. You have to sense how you shall work, as you see painted above.

Item: When you have set off with the graper, then throw the bridle on the hook which you shall have in front of you on your belt. Make sure to be properly in command of the horse with your body. Go up with the lance over your head so that your right hand is in front and the left is behind. Hit him with the graper below on his gorget so you push him from the horse as stated above.

Item: When someone rushes to you and tries to push you from the horse with the graper as painted before, then throw your lance away and grasp with your right hand in his lance at the graper. Press it decisively away from you, as painted above. This is how you throw him instantly instead of being thrown by him.

89R Item So du Sichst das dir ainer in den pausch greift | vnd hatt dir dein pruch genomen So las im dein spiess | Reitt wol in In vnd greiff mit deiner Rechten hannd | vnden an sein helm truck wol von dir So wirfst du in | von dem gaul wie oben gmalt Stat das haist der | halb Sunenblick

89V Item So ainer auff dich Rendt vnd wil dir dein gaul stechen | So setz dein spiess wol vir dich das er die erd berier | vnd berier vnd[5] zeuch auff ain Seitten auss mit deinem | gaul so hast im den stich oder stoss abgesetzt wie oben | gmalt statt

90R Item So du abgesetzt hast wie vorstett So zeuch wol auff | die Seitten halt dein spiess vest So Rentt[6] er gechling | darub*er* So wurfst du in von dem gaul vber ain hauffen | wie oben gmalt Statt also hat der spies mit Seiner arwait | ain end wie wol mer arbait darin ist Sy ist aber nit | ale versengclich

90V Item vermerck gar eben wie Sich das schwertt anfacht | im kampff zu fechten zu Ross die ersten zwei hutten | zu Ross nimpt er die schopfhutt wie oben gmalt statt | So schau das du dein schwertt auff die tengk Seyten | werffest vber dein lingke hand das das schwertt ~~oder~~ | ob der hand Sey das haist die nebenhutt

91R Item vermerck die ander zwu hutt So du Sichst das er die | nebenhat hutt[7] auff Seiner Rechten Seitten vnd Senck dein[8] | Schwertt wol nider an sein Rechten pain So du sichst die | nebenhutt So gee auff mit deiner Rechten handt vber deinen | kopff Senck dein Ortt wol virsich

5 'berier vnd' erroneously written twice.

6 The second 't' is hard to make out and could also be the beginning of either an 'h' or a 'k', left unfinished.

7 JWA2, fol. 32r: 'neben hutt hatt'.

8 'Senckt sein' would correlate to the image and would be more in line with the rest of the text.

Item: When you see that someone is grasping your graper and has foiled your break, then let him have your lance. Ride firmly into him and grab with your right hand below on his helmet. Press well away from you, and you will throw him off the horse as painted above. This is called the half of the sun view.

Item: When someone races to you and wants to thrust against your horse, then put your lance carefully before you so that it touches the ground. Move out to one side with the horse, so you have set off his thrust or hit, as painted above.

Item: When you have set off as described before, then pull well to the side and hold on to your lance. In this way, he races abruptly over it and you will throw him off the horse over the top as in the painting above. Here ends the lance with its work. Nevertheless, more work belongs to it, although someone will not always fall.

Item: Observe diligently the beginning of how to fence with the sword in mounted combat. The first two guards on horseback: If he assumes the *Schopfhut* as painted above, then make sure that you fling your sword to the left side over your left hand so that the sword is above the hand. This is called the *Nebenhut*.

Item: Note the other two guards. When you observe that he is assuming the *Nebenhut* on his right side and lowering his sword well down to his right leg, then you see the *Nebenhut*. Go up with your right hand over your head, lower your point accurately against him and pretend to hit him into the face.

gem [!] Im thu als wellest | du im in das gsycht stossen wie Sy baid oben gmalt stond | das haist die nebenhutt vnd die oberhutt

91v Item So dir ainer auss der ober hutt zu dem gsicht stost | So gee auff mit deiner neben hutt an Sein schwertt | So werstu Im den stoss wie oben gmalt Statt | darauss gat vil arwaitt

92r Item So du mit deim schwertt aufgangen bist vnd er | zuckt vnd Rucktt sein ortt hindersich vnd nimptt | den zornhaw auff dich So gee wol auff mit deim schwertt | vnd fach den straich in das kreutz wie oben gmalt stat | darauss gett vil arwait

92v Item So du Sichst das ainer den zornhau auff dich schlag | en wil oder mit dem knopff stossen So ge auff mit | deiner nebenhutt vnd Setz in an an drei ortten | welches du wiltt das merck ~~du~~ In dem anreitten | das erst ansetzen vnd Sein Rechten Iechssen das ander | In Sein Rechten elenpogen in das gelenck des harnasch | das dritt in das gsicht das helms wie oben gmalt | Statt vnd ist der pest grund

93r Item vermerck So ainer auff dich Rendt vnd wil das | ansetzen nemen oder das schirtzen So wind wol auff | mit deiner Rechten hand vnden[9] Sein schwertt treib | wol in die höch wie oben gmalt Stett So magstu | arwaiten was du wilt

93v Item So du das schwertt in die höch gwunden hast so | fal mit deim knopf des schwertt vber Sein Rechte | hand zeuch wol an dich Senck dich wol auff dem | gaul So prichst du im den arn oder nimbst Im | das schwertt man kan das stuck So gruntlich nitt | mallen oder schreiben vnd merck wie oben gmalt | Statt So magstu wol in die arwait kumen

9 JWA2, fol. 34r: 'vnden an'.

As they are both painted above, they are called the *Nebenhut* and the *Oberhut*.

Item: When someone hits to your face from the *Oberhut*, then go up with your *Nebenhut* at his sword. In this way, you ward off his hit as it is painted above. Much useful work ensues from this.

Item: When you've gone up with your sword and he twitches and jerks his point backwards in order to make a *Zornhaw* to you, then go well up with your sword and catch the strike on the crossguard, as it is painted above. Much work can be made from this.

Item: When you see that someone wants to strike a *Zornhaw* onto you or to hit with the pommel, then go up with your *Nebenhut* and set on in three spots, whichever you like. While riding against him, consider the first setting on under his right armpit; the second one in his right elbow in the joint of the armour; the third one into the sight of the helmet, as in the painting above. This is the best foundation.

Item: Observe that when someone races at you and wants to employ the setting on or the curtailing, then wind well up with your right hand below on his sword. Urge it up high, as painted above. In this way, you can work whatever you like.

Item: When you have wound the sword up high, then fall with the pommel of your sword over his right hand. Pull strongly towards you, and get down well on the horse. That's how you break his arm or take his sword. One can't paint or describe this piece sufficiently thoroughly. Note how it is painted above and you can very certainly come to your work.

94R Item So dich ainer also hat vberfallen wöllen So wind dein | knopff in die höch vnd das örtt gegen der erden vnd wind | wol vber sein arm das schwertt lass huren in dem mitelhaw | vber dein kopff legs im an Sein halss vnd greiff mit deiner | tengken hand in dein schwertt truck wol nider zeuch an dich | So wirfst du in von dem gaul deinenn zaum wirff an den | hacken wie oben gmaltt Statt Das stuck haist bschlie-sung | bei dem hals

94V Item So du Sichst das dich ainer bschliessen wil bey dem hals | So greiff mit deiner tengken hand an Sein Rechte Stoss | wol vbersich wend dich vmb ker im den Rugken als | wölest fliechen vnd schlach mit deinem schwertt nach | seinem tengke pain wie oben gmalt Statt ~~hau~~ das | haist der taschenhau

95R Item So du Sichst das Sich ainer vmbkertt in die flucht | So lass in nit vmkeren Eil in In stoss in aus der oberhutt | nach dem gsicht greiff mit deiner tengkhen hand auff sein | Rechte truck im sein hand vnd schwertt nider als da gmalt | Statt wirff den zam an den hagken fir dich den du haben | Solt vor dir

95V Item So ainer In dich Reitt vnd wil dir nach dem kopff[10] | Schlagen So darfstu nit alweg zucken var mit deine[r] | Rechten hand in seim rechten Elenpogen Reit wol In In | truck in virsich wol von dir wie da gmaltt Statt | So wirfstu in von dem gaul in ~~s~~ besser sicherhait so | Schlach mit deinem fuss vnden an Sein rechte pain | vnden an den sporn heb wol vbersich So hilfst du Im | zu dem wurff

96R Item So du Sichst das dich ainer also werffen wil oder mit | dem knopff stossen oder ansetzen wil was arwait er auff | dich treibtt So Reit wol In In vnd greiff mit deiner | Rechten hand zwischen Sein tengke hand durch den zam | var Im Im das gsycht zem*m* in In die

10 JWA2, fol. 36v: 'knopff'.

Item: When someone intends to assault you in such a way, then wind your pommel up high so that the point shows to the ground. Wind well over his arm. Hasten with the sword into a *Mitelhaw* over your head. Place it on his neck and grasp with your left hand in your sword. Press thoroughly down and pull towards you. In this way, you throw him off the horse. Throw your bridle on the hook, as in the painting above. This piece is called enclosing at the neck.

Item: If you see that someone wants to enclose you at the neck, then grasp with your left hand at his right. Push strongly upwards, turn around and turn your back to him as if you wanted to flee. Hit with your sword to his left leg as painted above. This is called the *Taschenhaw*.

Item: When you see that someone turns around in order to flee, then don't let him turn around. Rush into him and hit him to the face from the *Oberhut*. Grasp with your left hand on his right and press his hand and sword down as it is painted there. Throw the bridle forward on the hook which you shall have before you.

Item: When someone rides to you and wants to strike to your head, then you must never jerk. Go with your right hand in his right elbow and ride well into him. Press him forwards and strongly away from you as it is painted there. In this way, you throw him off the horse. In order to be even more secure, hit with your foot down at his right leg, below at the spurs. Lift him well up, so you allow him to fall.

Item: When you see that someone wants to throw you in this way, to hit with the pommel or to set on (or whatever work he intends against you), then ride right into him and grasp with your right hand between his left hand through the bridle.

höch bey Seiner | Schollern das haist der Sunenblick als oben gmaltt | Statt

96v Item das stuck merck gar eben wie es oben gmalt Statt es geitt | nit alweg ob es dir kumpt So nims kumpt es dir nit so lass | vnderwegen du wirst verkurtzt Sichstu aber das Sein gaul | gern in die höch gatt so schau das du nachent bei Seim kopff | des gauls Seiest vnd greiff vnd Sein kun zwisches[11] zamss be | greiff baid stangen oder zam wind wol in die höch mit deiner | Rechte hand So wirfstu den gaul vnd den man vber ain | hauffen du must gar eben auffmercken ob du es bekumen | magst oder nit wan du es begreifst So zeuch in auff ain | Seiten wol virsich

97r Item So ainer in dich rent vnd greift dir nach dem zam So | greiff mit deinem Rechten arm auff sein tengke achssel vnd | Schau das du in bei dem hals vnd helm vnden wol fasest | vnd Reit wol In In vnd mit deim Rechten fuss kum vnden | an sein Rechten fuss mit dem sporn wol hindersich vnd truck | oben wol vyr[12] dir So wirfstu in von dem gaul oder legst in | vir dich auff den Sattelpogen wie oben gmalt Statt das | Stuck haist die pressen

97v Item vermerck den pruch auff voder stuck So er die | pressen auff dich gnomen hat So var mit deinem Recht*en* | arm vir Sich vber Sein Rechten vnd greiff mit deiner | tengke hand vber Sein tengke das du baid baid zam zu samen | magst bringen[13] in die hand magst du es nit bekumen | So halt die tengken hand vest vnd Reit virsich ain | wenig auff die Seitten So wirfst du in von dem gaul | wie oben gmalt Statt ist vast ain gutter pruch

11 JWA2, fol. 37v: 'zwischen des'.

12 Corrected from 'von'.

13 JWA2, fol. 38v: 'mugst zu samen bringen'.

Go into his face and lift him up at his sallet. This is called the sun view, as it is painted above.

Item: Consider this piece thoroughly as it is painted above. It doesn't always work. If it is possible, make it; if it is not possible, leave it if you fall short on your way. But if you see that his horse tends to go up, then make sure to get close to his horse's head. Grab under its chin between the bridle and grasp both shanks or the bridle. Wind strongly high up with your right hand, so you throw the horse and the rider over the top. You need to take care whether you can accomplish this or not. When you grasp it, then pull him decisively forwards and to one side.

Item: When someone races into you and grabs for your bridle, then grasp with your right arm on his left shoulder and make sure that you safely hold him below at the neck and the helmet. Ride right into him and come with your right foot below at his right foot, with the spurs just backwards. Press above well forward, so you throw him off the horse or you lay him in front of you on the pommel, as painted above. This piece is called the pressing.

Item: Note the break against the previous piece. When he has undertaken the pressing against you, then go with your right arm forward over his right and grasp with your left hand over his left so that you manage to bring both bridles together in the hand. If you can't do this, hold on to the left hand and ride ahead a little to the side. In this way you throw him off the horse, as painted above. This is an excellent break.

98R Item So es in die flucht kumptt vnd ainer greift Dir hinden | in dein goler oder auff dein achssel haist auch die pressen ge | mit deiner lingke hand vnden durch Sein Rechtte arm vnd | wind oben vber sein Rechte mit deiner lingke hand buck dich | wol nider vnd haltt dein zam in deiner Rechte hand so prichst | du im dem arm oder wirft in von dem gaul wie oben gmalt | Stett es hatt gar gute arwait du must schauen das du wol | gricht Seyest in der kunst

98V Item vermerck den pruch auff das voder so dir ainer vber | windt mit Seim tengken vber sein[14] rechte arm vnden an | dich vnd fal bald oben vber sein tengke nachend bey den yechs*en* | leger Dich wol vir sich zeuch auss auff die Seiten das | kan man nit machen vnd merck wie der pruch oben | gmalt Stett So wirfstu in firsich auff den gaull

99R Item mer ain gutter bruch vnd stuck So ainer in dich Rendtt | vnd nimpt die pressen mit seiner Rechte hand auff dein tengke | achssel vnd wil die halb zwingen nemen vnd wil dich vber | Sich treiben in den wurff hindersich So fall bald mitt | deim Rechten arm oben vber Sein Rechte vnd begreiff sein | tengke bey dem zaum zeuch wol an dich wie da gmalt | Statt So wirfstu In

99V Item vermerckh den pruch oben darauff dan es statt gar | Klerlich nach ein ander So er dir den zaum begriffen hatt | mit seiner Rechten So fal mit deiner tengke auff sein Rechte | hand so er hindersich zie-chen wil So halt in vest vnd | wind mit dein Rechten arm vber sein baid arm[15] In den | schopff oder men deines gauls wie oben gmalt stat zeuch | wol auff ain[16] seitten von Im So wirfstu in von dem | gaul

14 Should be 'dein'

15 JWA2, fol. 40v: 'greiff'.

16 JWA2, fol. 40v has 'die' instead of 'ain'.

Item: When it happens that you flee and someone grasps your gorget from behind or on your shoulder, this is also called pressing. Go below with your left hand through his right arm and wind above over his right with your left hand. Bend well down and hold your bridle in your right hand. In this way, you break his arm or you throw him off the horse, as it is painted above. This one contains excellent work, and you need to take care to be well versed in the art.

Item: Note the break against the previous one. When someone winds with his left arm over your right arm [pull your right arm][17] below towards you, then fall instantly over his left, close to the armpit. Place yourself firmly in front and pull out to the side, which is impossible to paint. Note how the break which is painted above is done. That's how you throw him off[18] the horse in front.

Item: Another efficient break and piece. When someone races to you and employs the pressing with his right hand on your left shoulder and tries to apply half a constraining in order to force you up for a throw backwards, then immediately fall with your right arm above over his right and grasp his left at the bridle. Pull strongly towards you as painted above. That's how you throw him.

Item: Note above the break against this, for it is explained very clearly, one after another. When he has grasped your bridle with his right hand, then fall with your left on his right hand. As soon as he tries to pull it back, hold him tight and wind with your right arm over both of his arms. Grab the tuft or the mane of your horse as painted above and pull strongly to one side away from him. This is how you throw him off the horse.

17 Complemented according to O, fol. 165r.

18 Manuscript O makes it clear that the opponent is not thrown on but rather off the horse.

100R [19]Item vermerck das stuck das haist das prust stuckh vnd hatt drei pruch | So dir ainer nach der prust greift vnd wil dich auss dem sattel heben | So fal mit deinem Rechten arm vber sein Rechte in den zaum zu dein | er tengken hand So hastu im den pruch gethon

Item vermerck den andern pruch so du empfindst das er sich be gertt | vnd wil dich von dem gaul werffen in der zeit So ledig dein steg | raiff vnd halt dich vest an Im vnd Sitz von deinem gaul hin der | In wie oben gmalt statt So fierst du in wo du hin wiltt du | must in mit[20] baiden henden vmfachen in seinem zaum greiff | das haist die gfengknuss

100V Item so ainer auff dich eilt vnd pegertt dein so er auff | dich schlecht vnd du magst zu kainer wer nit kumen | So ge auff mit deiner Rechten hand an Sein schwertt greiff | virsich in sein schallern oder eisenhutt truck wol hind | ersich vnd Reitt wol virsich als da gmalt Statt das | Stuck haist der Sunenblick So wirfstu in von dem | gaul oder prichst im den hals

101R Item So einer auff dich eilt vnd kanst zu wer komen so schaw | das du das anwerffen wol bey dir habst wie dich der maister | lerntt wen die stuck sollen verborgen Sein man schreibtt | Sy nit gruntlich das an werffen wo du in trifst Reitt | nun von Im So wirfst du in von dem gaul wie oben | gmalt Statt

19 Paulus Hector Mair, previous owner of JWA2 in the 16th century, has left a curious remark on the corresponding page, fol. 41r: 'das stuckh Ist nit Miglich zu wegen zu bringen.' (This piece is impossible to make.)

20 JWA2, fol. 41r: 'deine'.

Item: Note the piece which is called the chest piece that has three breaks. When someone grasps for your chest and tries to lift you out of the saddle, then fall with your right arm over his right in the bridle to your left hand. In this way, you have achieved a break.
Item: Note the second break. When you sense that he has the desire to throw you off the horse at this time, then get rid of your stirrups. Hold on to him, leave your horse and sit behind him, as it is painted above. In this way, you can lead him wherever you want. You have to embrace him with both arms and grasp his bridle. This is called the apprehension.

Item: If someone rushes to you and seeks [to assault] you, and as soon as he strikes while you were unable to get to your weapons, then go up with your right hand at his sword. Grab in front at his sallet or iron hat, press strongly backwards and ride decidedly forward as painted above. This way you throw him off the horse or break his neck.

Item: If someone rushes to you and you are [un]able to get to your weapons, then make sure that you have the throwing on at your disposal, as the master has taught you. These pieces shall remain secret and one doesn't write them down explicitly. The throwing on: when you have hit him, ride away from him, so you throw him off the horse, as painted above.

101V Item vermerck das stuckh das gatt am ersten es gatt in | der mit vnd gat zu letst wen es dir komen mag seiest | wo du wellest in der arbait in treng vnd schlachen ist | es vast gutt vnd Im streitten vnd anderst schaw | das du bhend begreiffest oben in dem ~~schp~~ schopffraiff | ob las im in am halss hangen So ist er ain halber | man So magstu arwaiten was du wilt Du vindest | vor vnd nach vnd haist der schopfgriff wie oben | gmalt ist

Item Jörg wilhams hutters kunst zu augspurg etc

102R Wen ainer also auff dich Reit mit dem spiess wie oben gmalt statt | So schirtz wol vbersich in die höch so setzt im sein stich ab vnd | wie du hernach geschrieben vindst

So du im also abgesetzt hast lug das du gwis mit ar | wait seiest wind oben vber sein spies mit dem schwertt | oder was du vierst so nimst im sein spies oder ein schwertt | des selben gleich*en* oder was er vir ain wer viertt

Item: Note this piece which can be made in the beginning, in the middle and in the end, whenever it suits you, wherever you are in your work, in a melee or in battle. It is very efficient, in quarrels or else. Make sure to nimbly grasp the browband above. Let it hang at the neck, so he is only half a man and you can work whatever you want and which you'll find before and after. This is called the *Schopfgriff*, as in the paining above.

Item: The art of Jörg Wilhalm, hatter in Augsburg

When someone rides against you in the manner as it's painted above, then bolt well up high. This is how you set off his thrust, as you'll find written hereafter.

When you have set off, observe that you are secure in your work. Wind above over his lance with the sword or whatever you're carrying. In this same way, you take his lance or a sword or whatever weapon he's carrying.

APPENDICES

GLOSSARY

Abschneiden (*cutting off*): Using the edge of the sword in order to apply cuts to the opponent's openings (→*Blöss*). [32r, 33r, 34r, 44v, 45r]

Absetzen (*setting off*): A move that is suited for simultaneously attacking the opponent and defending oneself: From a displacement (→*Versatzung*), a thrust is the direct follow-up action. [29r, 29v, 30r, 30v, 31r, 13v, 43v, 44v, 45r, 45v, 50r, 85v, 86r, 87r, 87v, 89v, 90r, 102r]

after: →*Nach*.

Alber: A guard (→*Hut*) with the point (→*Ort*) towards the ground. The origin of the term remains unclear, since meanings encompass the silly fool (*alberner Narr*) and the poplar tree (*Alberbaum*). [12r, 23r, 43r, 43v, 44r, 46v, 47r, 47v]

altter Schnidt/alter Schnitt (*old cut*): A special cut through the opponent's arm. [17r, 26r, 34v, 35r, 44r, 49v]

Ansetzen (*settting on*): Bringing the point (→*Ort*) to one of the opponent's openings (→*Blöss*). [48r, 49v, 50v, 62v, 63r, 64r, 64v, 65r, 65v, 68r, 75r, 75v, 78v, 79v, 92v, 93r]

Arbait/Arwait, arbaiten/arwaiten (*work/working*): Attack; threatening the opponent. [2v, 13r, 26r, 27v, 43r, 43v, 44r, 44v, 48r, 48v, 49r, 49v, 50r, 61v, 62r, 62v, 63v, 65r, 65v, 66v, 71v, 72v, 74r, 74v, 76r, 76v, 78r, 78v, 89r, 80v, 85r, 85v, 86r, 87v, 90r, 91v, 92r, 93r, 93v, 96r, 98r, 101v, 102r]

armoured hand: →*gewappe[n]t hand*.

Band (*bind*): The contact of two weapons. [48v, 49r, 49v, 50v]

before: →*Vor*.

bevor: →*Partt*.

bind: → *Band.*

Blöss *(opening)*: A part of the body that is open to an attack. In armoured combat (→ *Harnasch*), this refers to those spots that are left unprotected by the steel plates and lames of the armour. In unarmoured combat, the body is divided into two upper and two lower openings, on the left and right sides. [9r, 9v, 11r, 11v, 36r, 43v, 45r, 48r, 49r, 49v, 50r, 50v, 82v]

break, breaking: → *Bruch.*

breaking window: → *Brechfenster.*

brechen: → *Bruch.*

Brechfenster *(breaking window)*: Misspelling of → *Sprechfenster.* [37v]

Bruch/Pruch *(break)*: Counter, or thwarting of an attack. [5v, 6r, 9r, 11r, 11v, 14r, 15r, 15v, 18r, 18v, 19r, 21r, 21v, 31r, 42r, 43v, 44v, 45r, 46r, 46v, 47v, 50r, 51r, 63r, 65r, 66r, 68r, 68v, 69v, 71v, 74r, 77r, 78r, 79v, 89r, 97v, 98v, 99r, 99v, 100r]

buffalo: → *Büffel.*

Büffel *(buffalo)*: A fencer who exploits his physical strength or superiority rather than putting much emphasis on technical finesse. [20r, 44r, 49r]

changer: → *Wechsel.*

changing through: → *Durchwechseln.*

clamper: → *Klemmer.*

clashing: → *glitzen.*

curtailing: → *Schirtzen.*

cutting off: → *Abschneiden.*

displacement, displacing: → *Versatzung.*

Dup[p]lieren *(doubling)*: An immediate follow-up strike from the bind (→ *Band*) that seems to double a first attack. [11r, 11v, 43v, 48r, 48v, 49v]

Durchlauffen *(running through)*: Wrestling techniques used to topple those opponents who like to run in with their arms high. [7v, 13v, 43v, 44v, 49v]

Durchwechseln *(changing through)*: Out of the bind (→*Band*), the point (→*Ort*) of the weapon is directed underneath the opponent's weapon in a semicircle in order to attack a new opening (→*Blöss*). [10r, 10v, 39r, 43r, 43v, 44r, 44v, 46r, 49v, 50r, 50v, 86v]

Einhorn *(unicorn)*: A guard (→*Hut*). [40v]

Eisen/Eysen *(iron)*: The head of the spear or lance. One of the two points (→*Ort*) of the weapon, the other being the butt where the →*Pausch* can be found. [85r, 86v, 87r, 87v]

Eisin portt/Eisnin wine portt/Eiserne Pforte *(iron gate)*: A guard (→*Hut*). The etymology of 'wine' is unclear and possibly only due to an error on the scribe's side. [12r, 12v, 48v]

Empfinden *(sensing)*: The sensation in the hand – or the hands – of whether the opponent is soft (→*waich*) or hard (→*hart*) in the bind (→*Band*) – also called *Fühlen* (feeling). As soon as one has felt this, various options appear for how to continue the fight (see also →*Indes*). [48v, 49r, 71r, 73r, 78r, 87v]

feeling: →*Empfinden.*

failer: →*Veler.*

Fehler: →*Veler.*

flat: →*Flech.*

Flech *(flat)*: The flat, broad side of the sword's blade. [35v, 43v, 44v]

Fullen *(feeling)*: →*Empfinden.*

Gerist *(lance rest)*: A support for holding the lance in mounted combat that is usually bolted or fastened onto the breastplate of the armour (→*Harnasch*). [85r, 87r]

gewappe[n]t hand *(armoured hand)*: A way of holding the sword in which the range of the sword is shortened by grabbing the middle of the blade with the left hand, in order to more powerfully and accurately make a thrust into the opponent's → *Blöss*. This occurs mostly in armoured combat when the hand is armoured with a gauntlet. However, this stance can be assumed in unarmoured combat as well. [18r, 28r, 28v, 41r, 79v]

Ghiltz/Ghultz *(hilt)*: The hilt of the sword, which consists of the crossguard, grip and pommel. Occasionally the crossguard alone or even a single quillon is called *Ghultz*. [46v, 47r, 50r, 65r, 74r, 76r, 78r]

glitzen/glutzen (*clashing*): The moment when the swords meet; literally: to flash or to glisten. [10v, 27v, 43v, 44v, 49r, 49v]

gloss: The verbose explanation of the very brief and often incomprehensible verses by Master Johannes → *Liechtenauer*, called 'the → *text*'.

going over: → *Ubergen*.

Goler (*gorget*): Neck protection that encloses the entire neck, unlike a bevor (→ *Partt*), which only protects the front. [98r]

graper: → *Pausch*. [87v, 88r, 88v, 89r]

guard: → *Hut*.

Haftung (*pinning*): Setting (→ *Ansetzen*) the point on certain parts of the armour. [62v, 63r, 63v, 64r, 64v, 65r, 65v, 68r, 75r, 75v, 78v, 79r, 79v, 81r, 82v]

Halbhaw: A technique that is only mentioned but not explained. [45r]

hanging: → *Hengen*.

hard: → *hart*.

Harnasch (*harness*): Suit of armour, plate protection with some additional mail elements to cover the remaining openings (→ *Blöss*). [61r, 71r, 79r, 92v]

hart (*hard*): Term for the opponent's lack of pliability in the bind (→ *Band*); see also → *waich*. [6v, 26r, 27v, 29r, 36v, 43r, 44r, 44v, 45r, 46r, 48r, 48v, 49r, 49v]

Haw: Strike.

Hend trucken (*pressing the hands*): The technique begins with a lower cut (→ *Schnitt*) of the sword to the hands of the opponent that is converted around the hands into an upper cut in one fluent motion. [35v, 36r, 43r, 44v]

Hengen (*hanging*): Those positions with the sword that threaten the opponent with the point (→ *Ort*). The upper hangings are identical to the → *Ochs*, the lower ones to the → *Pflug*. [3r, 3v, 14r, 25r, 36v, 37r, 43r, 44v, 45r, 45v, 51r]

hol Versatzung (*hollow displacement*): A technique that is mentioned but not explained. [51r]

hollow displacement: → *hol Versatzung.*

Hut (*guard*): A protective posture from which one can execute attacks; often used synonymously with → *Leger*. [12r, 14v, 17v, 46r, 46v, 47r, 47v, 49r, 50r, 50v, 61r, 61v, 62r, 85r, 87r, 90v, 91r]

Indes: Tactical situation in the fight deriving from the bind of the weapons (see also → *Band*). The terms instantly or simultaneously are only vague equivalents; it might be called a 'decisive moment', in which decisions about the further course of the fight are made. In the *Indes*, feeling (→ *Empfinden*) is of the utmost importance since one has to sense in what way the opponent has bound at one's weapon – either in a soft or hard way (→ *waich*, → *hart*). [2v, 5r, 26r, 43r, 44r, 48v, 49r, 49v]

iron: → *Eisen.*

Kampff: (Judicial) duel in armour (→ *Harnasch*); executed in the barriers – an enclosed fighting area – in front of an audience and under the supervision of judges and other aides. [61r, 80r, 81v, 82r, 83v, 84r, 85r, 90v]

Klemer (*clamper*): A technique in armoured combat in which the opponent's hand gets clamped between the swords. [67r, 67v]

Knopff (*pommel*): The back end of the sword, part of the hilt (→ *Ghültz*); counterweight for the blade. [27r, 44v, 46v, 47r, 64r, 65r, 66v, 67v, 74r, 75v, 76r, 76v, 78r, 92v, 93v, 94r, 96r]

Kreizhaw: A technique that is mentioned but not explained. [45r]

Krieg/edel Krieg (*war/noble war*): A fight situation from the bind (→ *Band*). In constant contact with the opponent's blade, one has to wind (→ *Winden*) in order to reach an opening (→ *Blöss*) with the point (→ *Ort*). [5r, 5v, 6r, 6v, 7r, 7v, 8r, 8v, 10v, 17r, 17v, 38r, 43r, 43v]

Krieghaw: A technique that is mentioned but not explained. [45r]

Kron (*crown*): A defensive position with the sword. [12v, 20v, 22r, 22v, 41r, 41v, 42r, 44r, 45r, 46v, 47v]

Kronhaw: A technique that is mentioned but not explained. [45r]

Krump, Krumphaw (*crooked strike*): One of Johannes →*Liechtenauer's* five secret strikes that follows a somewhat crooked trajectory to the opponent's hands or to his blade. [6v, 10r, 10v, 11v, 43r, 43v, 45r, 47v]

kurtze Schneid (*short edge*): The edge that faces towards the wrist; it is however just as long as the long edge (→*lange Schneid*) on a two-edged sword. [22v, 29r, 29v, 45v, 46v, 47r, 47v, 48r, 50r]

kurtzes Schwertt (*short sword*): This does not signify a short weapon, but rather a stance with the sword for combat in harness, often used synonymously for →*gewappe[n]t hand.* [37r]

langer Ort (*long point*): A guard (→*Hut*) with the point (→*Ort*) directed towards the opponent with long, outstretched arms. [30r, 30v, 31r, 31v, 38r, 46r, 46v, 48r, 50v]

lange Schneid (*long edge*): The edge of a bladed weapon that faces towards the finger joints; s. a. →*kurtze Schneid.* [3r, 8r, 32v, 45v, 47r, 48r, 48v, 49v, 50r]

Leger (*position*): →*Hut.*

Liechtenauer, Johannes: A fencing master from the 14th/15th centuries whose biographical data have not survived. His teachings in the form of rhymed verses are the basis for a major part of the German fencing treatises up to the end of the 16th century. His œuvre encompasses fighting in armour and on horseback, as well as unarmoured fencing with the sword. The latter in particular established many specific terms for strikes, thrusts, and additional fighting actions.

long point: →*langer Ort.*

lower cut: →*Unterschnitt.*

lower strike: →*Unterhaw.*

Messer: A single-edged weapon (literally a knife) with only a short back edge. The blade can be straight or moderately curved. Characteristic are the riveted handle scales as well as a riveted cross guard, where the rivet is prolonged on the outside to form a guard spike for additional hand protection.

Mitelhaw (*middle strike*): A horizontal strike. [94r]

Mordschlag (*murder strike*): A strike with the pommel (→*Knopff*) of the sword, with both hands at the blade, in order to conquer the opponent in armoured combat (→*Harnasch*). The sword is used like a club in order to cause damage to the joints or other parts of the armour through the heavy impact. [64r, 64v]

murder strike: →*Mordschlag*.

Muttieren (*mutating*): An attack is transformed from a strike (→*Haw*) in the bind (→*Band*) into a thrust, usually to a lower opening (→*Blöss*). [11r, 11v, 43v, 49v]

Nach (*after*): A situation in a fight when one has the disadvantage. With skilful attacking (→*arbaiten*) with the point (→*Ort*), however, one can gain the advantageous →*Vor* in order to threaten the opponent's openings (→*Blöss*). [2v, 5r, 43r]

Nachraisen (*travelling after*): The trajectory of one's own weapon follows the missed or failed strike of the opponent in the same direction in order to establish a threat or even a hit. [26r, 26v, 34r, 43r, 44r, 45r, 48r, 48v, 49r, 49v]

Nebenhut (*side guard*): A →*Hut* on horseback. The sword is held over the left hand. [90v, 91r, 92v]

Oberhaw (*upper strike*): A strike that attacks the upper openings (→*Blöss*). [3r, 3v, 4r, 4v, 10r, 10v, 13r, 13v, 43r, 45r]

Oberhut (*upper guard*): A →*Hut* on horseback with the sword held high over the head. [91r, 91v, 95r]

Oberschnitt (*upper cut*): A cut from above. [32r, 32v, 34r]

Ochs (*literally: ox*): A guard (→*Hut*) that protects the upper openings (→*Blöss*). [12r, 13r, 17r, 17v, 23r, 43v, 44r, 45r, 45v, 46r, 46v, 47r, 47v, 48r, 50r]

old cut: →*alter Schnitt*.

opening: →*Blöss*.

Ort (*point*): The tip of the weapon, usually that of a sword, →*Messer* or dagger. It can also be the blunt end of a staff or both ends of a lance (→*Eisen*, →*Pausch*). Occasionally synonymous with the → *thrust*.

[7v, 17r, 19v, 20r, 21r, 31r, 38r, 38v, 39r, 39v, 40r, 40v, 43v, 44r, 44v, 46r, 46v, 47r, 48r, 49v, 50r, 50v, 64r, 64v, 66r, 72v, 87v, 91r, 92r, 94r]

Partt (*bevor*): Chin protection of the →*Harnasch*, mostly worn in combination with a helmet called a sallet (→*Schaller*). [69v, 86v]

Pausch (*graper*): A ring made of leather or metal that can be put against the lance rest for stability while hitting the opponent. It is positioned close to the rear end of the spear or the lance – the butt, which is one of the two points (→*Ort*) of the weapon, the other being the →*Eisen*. [87v, 88r, 88v, 89r]

piece: →*Stück*.

Pflug (*plough*): A guard (→*Hut*) that protects the lower openings (→*Blöss*). [12r, 13r, 17r, 17v, 23r, 43v, 44r, 45r, 45v, 46r, 47r, 47v, 48r, 50r]

pinning: →*Haftung*.

point: →*Ort*.

pommel: 1. →*Knopff*. 2. The raised part at the front of a saddle.

position: →*Hut*.

Pressen: A technique on horseback in which one seeks close body contact with the opponent, generally in order to unhorse him. [97r, 97v, 98r, 99r]

pressing: →*Pressen*.

pressing the hands: →*Hend trucken*.

running over: →*Uberlauffen*.

running through: →*Durchlauffen*.

scales: →*Waage*.

Schaitelhaw, Schaitler: One of Johannes →*Liechtenauer's* five secret strikes (see →*Haw*) that aims high for the head. [12v, 22r, 22v, 43v, 44r, 45r, 46v, 47v]

Schaller/Scholler (*sallet*): A helmet with fixed or moveable visor that protects the skull. In order to protect the chin, an additional bevor (→*Partt*) could be worn. [96r, 100v]

Scher (*scissors*): A technique for armoured combat (→*Harnasch*) in which both swords are pushed crosswise over the opponent's head. [72v, 73v]

Schilcher, Schilcherhaw *(squinting strike)*: One of Johannes → *Liechtenauer's* five secret strikes, executed with the short edge (→ *kurtz Schneid*). [6v, 18r, 18v, 19r, 19v, 20r, 20v, 21r, 21v, 22r, 43r, 44r, 45r, 46r, 46v, 47v]

Schilthaw: Possibly a misreading of 'Schilhaw' (→ *Schilcher*). [44r]

Schirenprandt *(sword chafing)*: *Brant* is an old term for *sword*, and *schüren* is similar to the modern German *scheuern* (to chafe, to grind): The blades are in contact in a → *Band* (bind) and they move ever so slightly against each other. [74v]

Schirtzen *(curtailing)*: Technical term that is only mentioned once but without explanation. [93r]

Schnidt, Schnitt *(cut)*: A wound caused by a slice of the blade; the upper and the lower cut particularly aim for the opponent's arm, s. a. → *Hend trucken*.

Schopffhut *(tuft guard)*: A guard (→ *Hut*) on horseback with the point (→ *Ort*) of the sword directed toward the opponent. [90v]

Schopffgriff *(tuft grip)*: Technique on horseback in which one seizes the headpiece of the opponent's horse. [101v]

Schranckhut: Most commonly known as a guard (→ *Hut*) with the long sword held like a barrier or gate in front of the opponent, the point (→ *Ort*) towards the ground. Here, a position in armoured combat. [73r]

schwach, Schwech *(weak, weakness)*: 1. The half of the blade from the middle up to the point (→ *Ort*). 2. A position in which one presents the opponent a weak spot. [2v, 4r, 7r, 9v, 16v, 29v, 41r, 43r, 50r, 50v, 72v, 84r]

scissors: → *Scher*.

sensing: → *Empfinden*.

setting off: → *Absetzen*.

setting on: → *Ansetzen*.

shooting over: Quickly throwing the point forward, towards the opponent.

shoving away: → *Verschieben*.

soft: →*waich.*

speaking window: →*Sprechfenster.*

Sprechfenster (*speaking window*): A guard (→*Hut*). [37r, 40v, 45r]

starck, Sterck (*strong, strength*): 1. The half of the blade from the hilt (→*Ghultz*) up to the middle of the blade. 2. A position where one is or can be strong against the opponent. [2v, 4r, 7r, 9v, 13r, 16v, 25r, 29v, 33v, 37v, 41r, 43r, 43v, 50v, 72, 84r]

Stich: →*thrust.*

Straichen (*striking up*): A sweeping motion with the sword in which the point (→*Ort*) is struck upwards from below.

Striking up: →*Straichen.*

strong, strength: →*Stärke.*

Stuckh (*piece*): A technique or a sequence of techniques.

Sunnenblick (*sun view*): Wrestling grip on horseback that is used to turn the opponent's head upwards, in the direction of the sun. [89r, 96r, 100v]

sword chafing: →*Schirenprandt.*

Taschenhaw (*pouch strike*): An action with the sword on horseback in which one strikes to the opponent's left leg. [94v]

text: The largely cryptic verses of master Johannes →*Liechtenauer* which are explained with extensive →*glosses.*

thrust: Attack with the point of the sword (original: →*Stich*).

travelling after: →*Nachraisen.*

twitching: →*Zucken.*

Ubergen (*going over*): A technique in which one goes over the opponent's defence. [16r, 16v]

Uberlauffen (*running over*): The smaller reach of attacks towards the lower openings (→*Blöss*) is exploited with a strike or thrust to the upper ones. [5v, 19r, 27r, 27v, 28r, 28v, 43r, 44v, 45r, 49v]

Underhaw (*lower strike*): A strike that attacks the lower openings (→*Blöss*). [48r, 49r, 49v]

Underschnitt (*lower cut*): A cut from below. [33r]

upper cut: →*Oberschnitt.*

upper strike: → *Oberhaw.*

Veler (*failer*): Feint, mock assault. [3r, 3v, 4r, 13r, 43v]

Verhengen: → *Hengen.*

Verkerer: A technique that consists of a sequence of strikes and results in a bind (→ *Band*) from which one can thrust or run through (→ *Durchlauffen*). [7v, 13v]

Verschieben (*shoving away*): A technique that is only mentioned but not explained. [51r]

Versatzung, versetzen (*displacement, displacing*): Defensive action, parrying. The four displacements are techniques which contain a counterattack; they are more efficient than regular displacements in this way. [2r, 3r, 8r, 20r, 20v, 23r, 24r, 25r, 25v, 37r, 40r, 43v, 44r, 45r, 45v, 46r, 47r, 47v, 48r, 48v, 49r, 50r, 50v]

vom Tag (*from the day*): A guard (→ *Hut*) with an upright point (→ *Ort*). [12r, 14v, 23r, 43v, 47v, 48v]

Vor (*before*): A beneficial offensive fight situation in which one gains the advantage over an opponent by pre-empting his attack. [2v, 5r, 43r, 46v]

Wage (*scales*): A stable position that is frequently assumed in wrestling, although not exclusively there. [113v, 21r, 25r, 44r, 84r]

waich (*soft*): Yielding in the bind (→ *Band*); see → *hart.* [6v, 26r, 27v, 29r, 29v, 33v, 26v, 43r, 44r, 44v, 45r, 46r, 48r, 48v, 49r, 49v, 50v]

war/noble war: → *Krieg.*

weak, weakness: → *Schwech.*

Wechsel (*changer*): A strike that trails a preceding one backwards in the same direction; usually an → *Underhaw* that follows an → *Oberhaw.* [20r, 44r]

Winden (*winding*): Literally, turning. From the bind (→ *Band*) and without leaving it, the point (→ *Ort*) is directed towards the opponent's openings (→ *Blöss*) by means of rotating the weapon in a circle. [3r, 4r, 5v, 6r, 7r, 7v, 17v, 29r, 37v, 40r, 43r, 45r, 45v, 46r, 49r, 50r, 65r, 66v, 68v, 78r, 102r]

winding: → *Winden.*

work: →*Arbeit*.

Zornhaw: One of Johannes →*Liechtenauer's* five secret strikes with the long edge (→*lange Schneide*) directly onto the opponent's blade. A thrust with the point (→*Ort*) is the immediate follow-up action (→*Zornort*). In mounted combat, there is a similar powerful strike. [3r, 3v, 4v, 43r, 45r, 92r, 92v]

Zornort: A follow-up action from the →*Zornhaw* in which the point is directed to the opening (→*Blöss*) of the head or the chest of the opponent. [4r, 4v, 43r, 45r]

Zucken (*twitching*): If the opponent strikes to the weapon instead of the openings (→*Blöss*), the technique of instantly pulling back one's weapon so that the strike misses the target, and immediately striking back in return. [44v]

Zuvechten: The situation in which the opponents can attack or threaten each other with their weapons. [46r, 46v, 48v, 49r, 49v, 50r, 50v]

Zwerch, Zwerchhaw: One of Johannes →*Liechtenauer's* five secret strikes that is struck roughly above the head, going horizontally to either the right or left opening (→*Blöss*) of the opponent. The thumb is placed against the underside flat of the blade. [3r, 13r, 13v, 14r, 14v, 15r, 15v, 16r, 16v, 29r, 36v, 37v, 43r, 43v, 45r, 47v, 48v]

Zwerchwechselhaw: A technique that is only mentioned but not explained. [45r]

CONCORDANCE OF THE IMAGES

The following tables offer an overview of every single image in our *Leithandschrift* JWM1 in the far left column. The corresponding folio numbers of the examined manuscripts are presented next to each other from left to right in alphabetical order according to where they are kept, i.e. Augsburg comes first, followed by Dresden and so forth, ending with Wolfenbüttel. In some cases, an unambiguous identification was not possible. When in doubt, folio numbers (occasionally even several ones) are labelled with a question mark. In some instances, there are two images on one page; they are marked with -a (above) or -b (below) respectively. Note that manuscripts apart from JWM1 may contain additional material in each section which is not shown in the tables. PMD, PMM and PMW are double volumes, and the folio numbers refer to the first volume in each case.

SWORD

JWM1	JWA1	PMD	GE	PMM	JWM2	P	PMW	JWW
2r	–	26r? 53r?	3v	22r? 49r?	97r	12r?	5r? 32r?	15r
2v	–	–	4v	–	97v	–	–	15v
3r	6r	–	5v	–	98r	12v	–	16r
3v	6v	–	6v	–	98v	–	–	16v
4r	3r	–	7v	–	99r	–	61r	17r
4v	3v	–	8v	–	99v	–	–	17v
5r	7r	–	9v	–	100r	–	–	18r
5v	7v	37v? 38r?	10v	33v? 34r?	100v	–	16v? 17r?	18v
6r	4r	39r?	11v	35r?	101r	14r	18r?	19r

JWM1	JWA1	PMD	GE	PMM	JWM2	P	PMW	JWW
6v	5r	–	13v	–	101v	14v	–	19v
7r	5v	–	14v	–	102r	15r	–	20r
7v	4v	33r? 50v?	12v	29r? 46v?	102v	13v 12r?	12r? 29v?	20v
8r	38r	–	(15v)[1]	–	103r	–	–	21r
8v	38v	–	16v	–	103v	19v	61v	21v
9r	39r	–	17v	–	104r	15v?	–	22r
9v	39v	–	18v	–	104v	16v?	–	22v
10r	40r	48r?	19v	44r?	105r	–	27r?	23r
10v	40v	53r?	20v	49r?	105v	–	32r?	23v
11r	8r	–	21v	–	106r	17v?	–	24r
11v	8v	–	22v	–	106v	16v	–	24v
12r	14r	58r mirrored	79v	54r mirrored	107r	29v	37r mirrored	–
12v	14v	25v mirrored	80v	21v mirrored	107v	27r	4v mirrored	–
13r	7r	–	23v	–	108r	18r	–	–
13v	7v	–	24v	–	108v	–	–	–
14r	9r	60v	25v	56v	109r	18v	39v	26r
14v	9v	–	26v	–	109v	20r	–	26v
15r	10r	70v mirrored?	27v	66v mirrored?	110r	21v	49v mirrored?	25r
15v	10v	–	28v	–	110v	22r	–	25v
16r	11r	47v?	29v	43v?	111r	–	26v?	–
16v	11v	–	30v	–	111v	24v	–	–
17r	12r	–	31v	–	112r	–	62r	–
17v	12v	32v	32v	28v	112v	29v	11v	–
18r	18r	–	37v	–	113r	–	–	–
18v	18v	–	38v	–	113v	–	–	–
19r	19r	–	39v	–	114r	25r	–	–
19v	19v	–	40v	–	114v	–	–	–
20r	16r	–	33v	–	115r	–	–	27r

1 The digital data provided by Glasgow Museums do not contain this page. I was not able to verify whether this is due to an omission in the digitisation process or whether a sheet is actually missing. The consecutive numbering of the techniques in fact indicates a gap between nos. 12 and 14 (the numbers do not correlate to the foliation).

JWM1	JWA1	PMD	GE	PMM	JWM2	P	PMW	JWW
20v	16v	–	34v	–	115v	24r?	–	27v
21r	17r	–	35v	–	116r	–	–	28r
21v	17v	–	36v	–	116v	–	–	28v
22r	24r	28r mirrored?	41v	24r mirrored?	117r	–	7r mirrored?	29r
22v	24v	–	42v	–	117v	26r	–	29v
23r	25r	–	43v	–	118r	–	–	30r
23v	26r	–	45v	–	118v	–	–	30v
24r	26v	–	46v	–	119r	–	–	31r
24v	35v	54v?	44v	50v?	119v	17r	33v?	31v
25r	30r	49v	47v	45v	120r	–	28v	32r
25v	30v	50r	48v	46r	120v	–	29r	32v
26r	31r	–	49v	–	120r	22v	–	33r
26v	31v	40r?	50v	36r?	120v	23v	19r?	33v
27r	32r	58v?	51v	54v?	121r	30v?	37v?	34r
27v	32v	–	52v	–	121v	31r	–	34v
28r	33r	–	53v	–	122r	–	–	35r
28v	33v	68v?	54v	64v?	122v	50r	47v?	35v
29r	34r	–	55v	–	123r	–	–	36r
29v	34v	–	56v	–	123v	–	–	36v
30r	35r	–	57v	–	124r	–	–	37r
30v	35v	–	58v	–	124v	–	–	37v
31r	36r	–	59v	–	125r	–	–	38r
31v	36v	–	60v	–	125v	–	–	38v
32r	37r	–	61v	–	126r	26v?	–	39r
32v	37v	–	62v	–	126v	–	–	39v
33r	27r	–	63v	–	127r	–	–	40r
33v	27v	–	64v	–	127v	44v	–	40v
34r	28r	–	65r	–	128r	44r	–	41r
34v	28v	–	66v	–	128v	–	–	41v
35r	29r	–	67v	–	129r	–	–	42r
35v	29v	–	68v	–	129v	–	–	42v
36r	20r	–	69v	–	130r	–	–	–
36v	20v	31v? 34r mirrored	70v	27v? 30r mirrored	130v	–	10v? 13r mirrored	–
37r	21r	–	71v	–	131r	–	62v	–
37v	21v	–	72v	–	131v	–	63r	–
38r	22r	–	73v	–	132r	34v	–	43r
38v	22v	–	74v	–	132v	35r	–	43v

JWM1	JWA1	PMD	GE	PMM	JWM2	P	PMW	JWW
39r	23r	–	75v	–	133r	46v	–	44r
39v	23v	–	76v	–	133v	36r	–	44v
40r	13r	–	77v	–	134r	56v?	–	45r
40v	13v	–	78v	–	134v	56r?	–	45v
41r	15r	76r	81v	72r	135r	27v	55r	46r
41v	15v	76v	82v	72v	135v	28r	55v	46v
42r	41r	77r	83v	73r	136r	28v	56r	47r

HARNESS I

JWM1	JWM2	P	JWW
52r-a	170r	–	80r
52r-b	170v	163r	80v
52v-a	157r	–	81r
52v-b	157v	–	81v
53r-a	158r	154r	82r
53r-b	158v	–	82v
53v-a	159r	–	83r
53v-b	159v	–	83v
54r-a	160r	–	84r
54r-b	160v	172r	84v
54v-a	171r	–	85r
54v-b	171v	–	85v
55r-a	161r	175r	86r
55r-b	161v	155v?	86v
55v-a	162r	–	87r
55v-b	162v	–	87v
56r-a	163r	–	88r
56r-b	163v	174r	88v
56v-a	164r	–	89r
56v-b	164v	–	89r
57r-a	165r	–	90r
57r-b	165v	–	90v
57v-a	166r	–	91r
57v-b	166v	–	91v
58r-a	167r	–	92r
58r-b	167v	–	92v
58v-a	169r	166v	93r
58v-b	169v	–	93v

MISCELLANEOUS

JWM1	JWM2	P	JWW	O
58ar-a	210r	–	94r	85r
58ar-b	210v	194v	94v	84v
58av-a	211r	194r	95r	–

HARNESS II

JWM1	JWA2	JWA3	PMD	PMM	JWM2	PMW	JWW	O	JSW	JSWO1	JSWO2
59r	–	fragm. 1	–	–	156r	–	–	83v	–	–	–
59v	–	fragm. 2	203r	240v	156v	214v	–	86v	–	–	–
60r	1r	fragm. 3	203r	–	168r	214v	–	87v	86r	89r	87r
60v	1v	fragm. 4	203v	242v	168v	216v	–	diff.	86v	89v	87v
61r	2r	3r	204r	243r	172r	217r	57r	89v	87r	90r	88r
61v	2v	3v	204v	243v	172v	217v	57v	90v	87v	90v	88v
62r	3r	4r	205r	244r	173r	218r	58r	91v	88r	91r	89r
62v	3v	4v	205v	244v	173v	218v	58v	92v	88v	91v	89v
63r	4r	5r	206r	245r	174r	219r	59r	93v	89r	92r	90r
63v	4v	5v	206v	245v	174v	219v	59v	94v	89v	92v	90v
64r	5r	6r	207r	246r	175r	220r	60r	95v	90r	93r	91r
64v	5v	6v	208r	247r	175v	221r	60v	96v	90v	93v	91v
65r	6r	7r	208v	247v	176r	221v	61r	97v	91r	94r	92r
65v	6v	7v	209v	248v	176v	222v	61v	98v	91v	94v	92v
66r	7r	8r	210v	249v	177r	223v	62r	99v	92r	95r	93r
66v	7v	8v	211r	250r	177v	224r	62v	100v	92v	95v	93v
67r	8r	9r	212r	251r	178r	225r	63r	101v	93r	96r	94r
67v	8v	9v	213r	252r	178v	226r	63v	102v	93v	96v	94v
68r	9r	10r	213v	252v	179r	226v	64r	103v	94r	97r	95r
68v	9v	10v	214r	253r	179v	227r	64v	104v	94v	97v	95v
69r	10r	11r	214v	253v	180r	227v	65r	105v	95r	98r	96r
69v	10v	11v	215v	254v	180v	228v	65v	106v	95v	98v	96v
70r	11r	12r	216r	255r	181r	229r	66r	107v	96r	99r	97r
70v	11v	12v	216v	255v	181v	229v	66v	108v	96v	99v	97v
71r	12r	13r	217r	256r	182r	230r	67r	109v	97r	100r	98r

JWM1	JWA2	JWA3	PMD	PMM	JWM2	PMW	JWW	O	JSW	JSWO1	JSWO2
71v	12v	13v	217v	256v	182v	230v	67v	110v	97v	100v	98v
72r	13r	14r	218v	257v	183r	231v	68r	111v	98r	101r	99r
72v	13v	14v	219r	258r	183v	232r	68v	112v	98v	101v	99v
73r	14r	15r	219v	258v	184r	232v	69r	113v	99r	102r	100r
73v	14v	15v	220v	259v	184v	233v	69v	114v	99v	102v	100v
74r	15r	16r	221v	260v	185r	234r	70r	115r	100r	103r	101r
74v	15v	16v	222v	261v	185v	235v	70v	116v	100v	103v	101v
75r	16r	17r	223r	262r	186r	236r	71r	117v	101r	104r	102r
75v	16v	17v	224r	263r	186v	237r	71v	118v	101v	104v	102v
76r	17r	18r	224v	263v	187r	237v	72r	119v	102r	105r	103r
76v	17v	18v	225r	264r	187v	238r	72v	120v	102v	105v	103v
77r	18r	19r	227v	266v	188r	240v	73r	121v	103r	106r	104r
77v	18v	19v	228r	267r	188v	241r	73v	122v	103v	106v	104v
78r	19r	20r	228v	267v	189r	241v	74r	123v	104r	107r	105r
78v	19v	20v	229r	268r	189v	242r	74v	124v	104v	107v	105v
79r	20r	21r	229v	268v	190r	242v	75r	125v	105r	108r	106r
79v	20v	21v	230r	269r	190v	243r	75v	126v	105v	108v	106v
80r	21r	22r	230v	269v	191r	243v	76r	127v	106r	109r	107r
80v	21v	22v	231r	270r	191v	244r	76v	128v	106v	109v	107v
81r	22r	23r	238r	280r	192r	254r	77r	129v	107r	110r	108r
81v	22v	23v	238v	280v	192v/ 193v	254v	77v	130v	107v	110v	108v
82r	23r	24r	239r	281r	193r	255r	78r	131v	108r	111r	109r
82v	23v	24v	239v	281v	194r	255v	79r	132v	108v	111v	109v
83r	24r	25r	240r	282r	194v	256r	79v	133v	109r	112r	110r
83v	24v	25v	240v	282v	195r	256v	96r	134v	109v	112v	110v
84r	25r	26r	241r	283r	195v	257r	96v	135v	110r	113r	111r

MOUNTED COMBAT

JWM1	JWA2	JWA3	PMD	PMM	JWM2	PMW	JWW	O	JSW	JSWO1	JSWO2	P
85r	26r	26v	276r	175v	138r	288v	97r	137v	112v/ 113r	114v/ 115r	113v/ 114r	–
85v	26v	27r	276v	176r	138v	289r	97v	138v	114v/ 115r	116v/ 117r	115v/ 116r	–
86r	27r	27v	277v	178r	139r	291r	98r	139v	116v/ 117r	118v/ 119r	117v/ 118r	–
86v	27v	29r	278v	179v	139v	292v	98v	140v	118v/ 119r	120v/ 121r	119v/ 120r	–

JWM1	JWA2	JWA3	PMD	PMM	JWM2	PMW	JWW	O	JSW	JSWO1	JSWO2	P
87r	28r	29v	279r	180r	140r	293r	99r	141v	120v/ 121r	122v/ 123r	121v/ 122r	134v
87v	28v	30r	279v	180v	140v	293v	99v	142v	122v/ 123r	124v/ 125r	123v/ 124r	–
88r	29r	30v	280r	181r	141r	294r	100r	143v	124v/ 125r	126v/ 127r	125v/ 126r	–
88v	29v	31r	280v	181v	141v	294v	100v	144v	126v/ 127r	128v/ 129r	127v/ 128r	–
89r	30r	31v	281r	182r	142r	295r	101r	145v	128v/ 129r	130v/ 131r	129v/ 130r	–
89v	30v	32r	281v	182v	142v	295v	101v	146v	130v/ 131r	132v/ 133r	131v/ 132r	–
90r	31r	32v	282r	183r	143r	296r	102r	147v	132v/ 133r	134v/ 135r	133v/ 134r	–
90v	31v	33r	282v	184r	143v	297r	102v	148v	148v/ 149r	150v/ 151r	149v/ 150r	–
91r	32r	33v	283r	184v	144r	297v	103r	149v	150v/ 151r	152v/ 153r	151v/ 152r	–
91v	32v	34r	283v	185r	144v	298r	103v	150v	152v/ 153r	154v/ 155r	153v/ 154r	137r
92r	33r	34v	284r	185v	145r	298v	104r	151v	154v/ 155r	156v/ 157r	155v/ 156r	–
92v	33v	35r	284v	186r	145v	299r	104v	152v	156v/ 157r	158v/ 159r	157v/ 158r	138r
93r	34r	35v	285r	186v	146r	299v	105r	153v	158v/ 159r	160v/ 161r	159v/ 160r	139v?
93v	34v	36r	285v	187r	146v	300r	105v	154v	160v/ 161r	162v/ 163r	161v/ 162r	–
94r	35r	36v	286r	188r	147r	301r	106r	155v	162v/ 163r	164v/ 165r	163v/ 164r	–
94v	35v	37r	286v	188v	147v	301v	106v	156v	164v/ 165r	166v/ 167r	165v/ 166r	200v/ 201r
95r	36r	37v	287r	189r	148r	302r	107r	157v	166v/ 167r	168v/ 169r	167v/ 168r	–
95v	36v	38r	287v	189v	148v	302v	107v	158v	168v/ 169r	170v/ 171r	169v/ 170r	–
96r	37r	38v	288r	190r	149r	303r	108r	159v	170v/ 171r	172v/ 173r	171v/ 172r	–
96v	37v	39r	288v	190v	149v	303v	108v	160v	172v/ 173r	174v/ 175r	173v/ 174r	–
97r	38r	39v	289r	191r	150r	304r	109r	161v	174v/ 175r	176v/ 177r	175v/ 176r	–

JWM1	JWA2	JWA3	PMD	PMM	JWM2	PMW	JWW	O	JSW	JSWO1	JSWO2	P
97v	38v	40r	289v	191v	150v	304v	109v	162v	176v/ 177r	178v/ 179r	177v/ 178r	141r?
98r	39r	40v	290r	192r	151r	305r	110r	163v	177v/ 178r	180v/ 181r	179v/ 180r	–
98v	39v	41r	290v	192v	151v	305v	110v	164v	179v/ 180r	182v/ 183r	181v/ 182r	–
99r	40r	41v	291r	193r	152r	306r	111r	165v	181v/ 182r	184v/ 185r	183v/ 184r	–
99v	40v	42r	291v	193v	152v	306v	111v	166v	183v/ 184r	186v/ 187r	185v/ 186r	–
100r	41r	42v	291r	194r	153r	307r	112r	167v	185v/ 186r	188v/ 189r	187v/ 188r	–
100v	41v	43r	291v	194v	153v	307v	112v	168v	187v/ 188r	190v/ 191r	189v/ 190r	–
101r	42r	43v	292r	195r	154r	308r	113r	169v	–	–	–	–
101v	42v	44r	292v	195v	154v	308v	113v	170v	191v/ 192r	194v/ 195r	193v/ 194r	–
102r-a	–	28r	277r	176v	155r	289v	114r	–	–	–	–	135v
102r-b	–	28v	278v	177r	155v	290r	114v	–	–	–	–	–

BIBLIOGRAPHY

MANUSCRIPTS

Augsburg, Universitätsbibliothek, Codex I.6.2°.2 (JWA1).
Augsburg, Universitätsbibliothek, Codex I.6.2°.3 (JWA2).
Augsburg, Universitätsbibliothek, Cod.I.6.2°.5 (SS).
Augsburg, Universitätsbibliothek, Cod.I.6.4°.2 (B).
Augsburg, Universitätsbibliothek, Codex I.6.4°.5 (JWA3).
Copenhagen, Det Kongelige Bibliotek, Thott 290 2° (HTK).
Dresden, Sächsische Landesbibliothek, Mscr. Dresd. C.93/94 (PMD).
Glasgow, R. L. Scott Collection, E.1939.65.341 (SE).
Glasgow, R. L. Scott Collection, E.1939.65.354 (GE).
Heidelberg, Universitätsbibliothek, Cod. Pal. Germ. 430 (LH).
Jena, Thüringer Universitäts- und Landesbibliothek, MS G.B.f.18.a (J).
Krakow, Biblioteka Jagiellonska, Ms. Germ. Quart. 16 (GK).
Krakow, Biblioteka Jagiellonska, Ms. Germ. Quart. 2020 (G).
Leeds, Royal Armouries Museum, I.33 (I).
Munich, Bayerische Staatsbibliothek, Cod. icon. 393 (PMM).
Munich, Bayerische Staatsbibliothek, Cgm 582 (LM).
Munich, Bayerische Staatsbibliothek, Cgm 3711 (JWM1).
Munich, Bayerische Staatsbibliothek, Cgm 3712 (JWM2).
New Haven, Yale Center for British Art, MS U860.F46 1450 (GN).
Paris, Musée National du Moyen Age, CL23842 (P).
Rome, Biblioteca dell'Academia Nazionale dei Lincei e Corsiniana, 44 A 8 (PD).
Vienna, Kunsthistorisches Museum, KK 5012 (PF).

Vienna, Kunsthistorisches Museum, KK 5013 (GW).
Vienna, Kunsthistorisches Museum, KK 5247 (JSW).
Vienna, Österreichische Nationalbibliothek, Cod. Vindob. 10825/6 (PMW).
Wolfenbüttel, Herzog August-Bibliothek, Cod. Guelf. 38.21 Aug. 2° (JWW).
Wolfenbüttel, Herzog August-Bibliothek, Cod. Guelf. 1.6.3 Aug. 2° (JSWO1).
Wolfenbüttel, Herzog August-Bibliothek, Cod Guelf. 78.2 Aug. 2° (GWO).
Wolfenbüttel, Herzog August-Bibliothek, Cod. Guelf. 5 Blank. (JSWO2).
Private property; formerly Gräflich Ortenburg'sche Bibliothek zu Tambach, Ortenburger Fechtbuch, olim Cart. Misc. S XVI (O).

SECONDARY SOURCES

CINATO 2016

FRANCK CINATO: Development, Diffusion and Reception of the 'Buckler Play': A Case Study of a Fighting Art in the Making. In: JAQUET/VERELST/DAWSON 2016.

FORGENG 2018

JEFFREY L. FORGENG: *The Medieval Art of Swordsmanship. Royal Armouries MS I.33*. Royal Armouries Museum, Leeds 2018.

HAGEDORN/WALCZAK 2015

DIERK HAGEDORN/BARTŁOMIEJ WALCZAK: *Gladiatoria – New Haven*. MS U860.F46 1450. VS-Books, Herne 2015.

HAGEDORN/WALCZAK 2018

DIERK HAGEDORN/BARTŁOMIEJ WALCZAK: *Gladiatoria – Medieval Armoured Combat*. Greenhill Books, Barnsley 2018.

HAGEDORN 2016

DIERK HAGEDORN: German Fechtbücher from the Middle Ages to the Renaissance. In: JAQUET/VERELST/DAWSON 2016.

HAGEDORN 2017

DIERK HAGEDORN: *Jude Lew – Das Fechtbuch*. VS-Books, Herne 2017.

HAGEDORN/AMBERGER 2020

DIERK HAGEDORN/J. CHRISTOPH AMBERGER: *Codex Amberger*. VS-Books, Herne 2020.

HILS 1985

HANS-PETER HILS: *Meister Johann Liechtenauers Kunst des langen Schwertes*. Peter Lang, Frankfurt am Main/Bern/New York 1985.

HILS 1987

HANS-PETER HILS: 'Gladiatoria'. Über drei Fechthandschriften aus der ersten Hälfte des 15. Jahrhunderts. In: *Codices manuscripti, Heft 1/2*. Brüder Hollinek, Vienna 1987.

JAQUET/VERELST/DAWSON 2016

DANIEL JAQUET/KARIN VERELST/TIMOTHY DAWSON (EDITORS): *Late Medieval and Early Modern Fight Books*. Brill, Leiden/Boston 2016.

KRAUSE/PFAFFENBICHLER 2017

STEFAN KRAUSE, MATTHIAS PFAFFENBICHLER (EDITORS): *Turnier: 1000 Jahre Ritterspiele*. Hirmer, Munich 2017.

LENG 2008

RAINER LENG (EDITOR), HELLA FRÜHMORGEN-VOSS, NORBERT H. OTT, ULRIKE BODEMANN, PETER SCHMIDT, CHRISTINE STÖLLINGER-LÖSER: *Katalog der deutschsprachigen illustrierten Handschriften des Mittelalters Band 4/2, Lfg. 1/2: 38. Fecht- und Ringbücher*. C. H. Beck, Munich 2008.

SCHMELLER 1866

JOHANN ANDREAS SCHMELLER: *Die deutschen Handschriften der K. Hof- und Staatsbibliothek zu München, zweiter Teil*. Palmsche Hofbuchhandlung, Munich 1866.

WALCZAK 2015

BARTŁOMIEJ WALCZAK: The Gladiatoria Group. An Overview. In: HAGEDORN/WALCZAK 2015.

WELLE 2014

RAINER WELLE: *... vnd mit der rechten faust ein mordstuck. Baumanns Fecht- und Ringkampfhandschrift*. Herbert Utz Verlag, Munich 2014.

WIERSCHIN 1965

MARTIN WIERSCHIN: *Meister Johann Liechtenauers Kunst des Fechtens*. C. H. Beck, Munich 1965.

WILHALM 2017

JÖRG WILHALM: *Die Fechtkunst von Jörg Wilhalm Huter*. Fines Mundi, Saarbrücken 2017.

ONLINE SOURCES

Wiktenauer

wiktenauer.com/wiki/Jörg_Wilhalm_Hutter

FamilySearch

familysearch.org

Wasserzeichen-Informationssystem

wasserzeichen-online.de

ABOUT THE AUTHOR

Dierk Hagedorn works for the IT subsidiary of a large German airline for numerous creative issues such as user experience, interface and corporate design. He is an instructor for the long sword at Hammaborg, a club for historical fencing in Hamburg; he is also a passionate wearer of armour. He has transcribed 30 of the approximately 80 surviving fight books in the German language and made them available on the website of his club or published them in book form. The current volume is his thirteenth edition of an old fight book, and the first time that he has teamed up with his two children, Helen and Henri, in order to accomplish this achievement.

ACKNOWLEDGEMENTS

When my publisher, Michael Leventhal, approached me for yet another fight book edition – a follow up to the two full-colour volumes about the *Gladiatoria* manuscript from New Haven and the famous Talhoffer codex from Munich – I didn't need to think excessively long. Years ago, I had transcribed other exemplars from Wilhalm's hand, while only the most complete and splendid one from Munich was still missing. Quickly, we came to terms. I'm happy that I have been able to follow my whims almost unhampered since I insist on being responsible for the design of my books myself. I'm grateful for this accommodation and confidence which not every author can enjoy.

I remain more than happy to work with the indispensable Elani Wales again, whose sharp eye always helps my foreign and whimsical English to shine at least a bit.

A lot of information needed to be gathered from institutions in Germany and Austria, and so I cannot thank the following lady and gentlemen enough for always being able to provide me with competent and quick answers to my sometimes complicated and extensive questions: Dr. Andreas Kosuch (Universitätsbibliothek Augsburg), Dr. Stefan Krause (Kunsthistorisches Museum Wien, Imperial Armoury) and Alexandra Schebesta (Herzog August-Bibliothek Wolfenbüttel).

I want to thank Alexander Kalywas and Christian Trosclair for their valuable insight; and a very special thank you is due to Arne Koets, who provided me with quite specific equine counsel.

The original manuscript was created in Augsburg in the Early Renaissance, so I thought it appropriate to use an Early Renaissance typeface. My initial choice of Charpentier Renaissance, created by Ingo Zimmermann, a typographer from – conveniently – Augsburg, proved to lack the essential small caps. Upon special request, he instantly agreed to upgrade that splendid font, which I found extremely helpful of him. I heartily recommend his work to anyone with a typographic sensibility.

Finally, I would like to thank my beloved wife Verena, who has no interest in fight books at all, but is always supportive and has given me two wonderful children. This time, however, I'm afraid I overdid it a bit by encouraging these children to cooperate with me. I hope I did not alienate them too much from their mother. Whom I truly love.